Introduction to Dynamics

Friedrich Pfeiffer · Thorsten Schindler

Introduction to Dynamics

Springer

Friedrich Pfeiffer
Institute of Applied Mechanics
Technische Universität München
Garching
Germany

Thorsten Schindler
Institute of Applied Mechanics
Technische Universität München
Garching
Germany

ISBN 978-3-662-52280-6 ISBN 978-3-662-46721-3 (eBook)
DOI 10.1007/978-3-662-46721-3

Springer Heidelberg New York Dordrecht London

Softcover reprint of the hardcover 1st edition 2015

Printed on acid-free paper

Springer-Verlag GmbH Berlin Heidelberg is part of Springer Science+Business Media
(www.springer.com)

Preface

As a fundamental science in both physics and engineering, mechanics deals with interactions of forces resulting in motion and deformation of material bodies. Mechanics serves in the world of physics and in that of engineering in a particular way, in spite of many increasing interdependencies. For physicists, machines and mechanisms are tools for cognition and research. For engineers, they are the objectives of research, according to a famous statement by the Frankfurt physicist and biologist Friedrich Dessauer. Physicists apply machines to support their questions to nature with the goal of new insights into the physical world. Engineers apply physical knowledge to support the realization process of their ideas and their intuition. Physics is an analytic science searching for answers to questions concerning the world around. Engineering is a synthetic science, where the physical and mathematical fundamentals play the role of reinsurance with respect to a working and efficiently operating machine. Engineering is an iterative science typically resulting in long development periods of its products. However, it is also iterative concerning fast improvements of an existing product or fast development of a new one. Every physical or mathematical science has to face these properties by formulating new specific methods, new practically approved algorithms up to new fundamentals, which are adaptable to new technological developments. This is also true for the field of mechanics.

With the increasing complexity of technical products, we need more sophisticated design methods. Therefore, we have to systematically apply all modern options of theoretical and experimental modeling to achieve optimal realization processes. Product development alone by intuitive design and experimental tests is unreasonably costly. Theoretical simulations replace more and more experiments, or at least they reduce tests by organizing them in a sensible way. This definitely requires good models, which must be close to reality. Realism needs awareness of the most important aspects, which finally decide the size and quality of models and theory.

This book gives an introduction to dynamics and tries to consider the above recommendations. Thereby, a compromise has to be found between basics and applications. The fundamental principles and laws of kinematics and kinetics are treated in a general way and can be applied to more complex cases. Furthermore, they are transparent and clear enough to be application-friendly. We start with these considerations at the very beginning. Afterwards, we treat a few selected problems of linear and nonlinear dynamics, and finally, we present a more phenomenologically oriented chapter on problems of vibration formation. Obviously such an introduction to dynamics can only present a selection of topics, which nevertheless should give a useful basis for stepping into more advanced problems of dynamics. The selection in this book represents the result of a regularly revised course, which has been and still is given for masters students at the Technische Universität München.

We owe many conceptional items together with some ideas of conceptual details to Kurt Magnus, who founded the chair B of Mechanics at the Mechanical Engineering Department of the Technische Universität München with a focus on dynamics in 1966. Friedrich Pfeiffer, his successor, continued the course's evolution by including his significant industrial experience considering mainly problems connected with theory and practice. Heinz Ulbrich adapted the course to include his view of mechanics, and finally, we have input from the younger generation represented by Thorsten Schindler who teaches dynamics to students. We thank Daniel Rixen, the present chair, for the possibility to elaborate this version of the book.

Generations of assistants and students have worked for this course, the one or other presenting the course him/herself thus helping to improve it. Our warmest regards to them, especially for continual discussions of contents and formulations. Thanks also to the former Teubner Verlag and to the Springer Publishing House for the excellent problem-free cooperation.

Munich, February 2015 *Friedrich Pfeiffer, Thorsten Schindler*

Contents

Chapter 1
Basics

1.1 Introduction

"Mechanics is the science of motion; we define as its task: to describe *completely* and in the *simplest possible manner* such motions as occur in nature." With respect to engineering we should complete this statement by "as occur in nature and in technology." This more than a hundred-year-old statement was made by KIRCHHOFF [34] and has lost neither its meaningfulness nor its assertion. Technical mechanics as a science must also be as simple as possible but conversely descriptively complete. If we consider motion as any kind of translation and rotation, even if only minimal as in the case of deformations, and also include the state of no motion (i.e., the state of rest), then motion describes mechanics as a whole. It comprises two fundamental aspects, that of geometry and kinematics describing positions and orientations, velocities and accelerations, and that of kinetics, describing the cause of motion. Regarding all possible interactions between material bodies or between material bodies and their environment, we consider those possibilities, which produce accelerations (or deformations) of these bodies. We call the driving magnitude of such interactions *forces*. Thus, the kinematics of bodies and their interaction with forces, their statics and kinetics, define mechanics.

From a physical point of view, classical mechanics might be a self-contained area of science, the laws of which can be derived completely in a deductive way. Classical mechanics can be reduced to some fundamental axioms if we know the laws of force interactions. Axioms cannot be proven, but their statements have to be in accordance with practical experience, without exception. Such a deductive approach is only possible for idealized models, which themselves represent a reduced picture of the real world [16]; for many technical applications, it is not possible.

1.2 Modeling

With respect to technical mechanics, the aspect of *modeling* becomes one of the most important issues of mapping real world problems. Technical mechanics is an

F. Pfeiffer and T. Schindler, *Introduction to Dynamics*,
DOI: 10.1007/978-3-662-46721-3_1

engineering science, which considers motion or deformations of technical systems. They generate loads on machines, mechanisms and structures, which must be known for their design. *Mechanical modeling* includes the replacement of a real machine, of real machine components or of real structures by certain basic elements. Considering mechanics, this concerns for example masses, springs, dampers and frictional elements, which according to the topology of a structure must be interconnected in a physically correct way, usually leading to certain types of constraints. This process requires a deep insight into the operational problems of a machine and a sound knowledge of practice on the one and of mechanical theories on the other side. The quality of modeling decides on time and costs during a product development process, at least to a certain amount. Good models not only lead to quicker solutions, but also to better transparency of the problem under consideration and with it to accelerated achievements for a technical problem.

What is a *good model*, or better, what is a *good mechanical model*? A mechanical model will be good, if the mathematical model based on it gives us information close to the reality, which is of special interest to us. We have to anticipate, that reality is known, that it might be measurable or at least that it might be precisely describable. Therefore a good model should help us to come to a deeper understanding of the technical problems involved and of the design ideas behind them. To produce only numbers and charts will be not enough, we would like to produce insight. Creating models has to keep that in mind.

How can we achieve a mechanical model? Usually we can assume, that every machine, machine component or structure offers some important operational functions, which are easy to describe and to model. With regard to mechanical systems, these are for example some idealized motion sequences or vibrations, some effects from kinematics and kinetics. We start with that. Looking a bit deeper into a structure, we might realize, that machines cannot be built in an ideal way, that we are confronted with disturbances, with "dirty effects", which in many cases cannot be modeled straightforwardly and the mechanics of which is often not understood. Exactly at that point, the typical work of an engineer starts, which possesses more an intuitive-empirical character than a scientific one, for example the question, what can be neglected. A good mechanical model is always a minimalist model, not smaller than necessary, but also not larger than adequate to the problem involved. Finding intuitively neglections, we may consider the geometric and kinematic situation, the order of magnitude of forces and torques or of physical work and energy. Establishing a good model always needs an iteration process, which leads us with every step to a better solution. In his famous lecture on "Clouds and Clocks" from 1965 [54], Karl Popper told, that iterations are not only characteristic features of every intellectual work, but that they lead also from step to step to a deeper insight to the problem and to new questions finally achieving a really innovative solution, which at the beginning of such a process could not be perceived:

- mechanical modeling (theoretically and/or experimentally),
- examination with respect to plausibility, comparisons with reality,
- adaptation and improvements of models.

Establishing a mechanical model, we have to watch some important aspects concerning mathematical and numerical modeling, that means the whole sequence of steps to a final solution of our mechanical model ideas:

- *discretization*: Can we compose our model only by rigid body elements or even by point masses or do we have to use elements with a continuum-mechanical character? How shall we model such nonrigid bodies?
- The character of expected motion: Does some basic motion exist or do we have a type of *reference motion*? Is there some state of rest? Is it possible to describe the motion as one with a (usually nonlinear) reference motion and small deviations from it? Can we linearize, completely or in parts?
- *coordinates*: How many degrees of freedom exist for our model? Can we find a set of coordinates, which corresponds directly to these degrees of freedom? If not, what sets of coordinates offer a formulation of constraints in a most simple way?
- *numerics*: What solution methods fit best to our problem, analytically (if any) or numerically? Can we put the mathematical formulation in a form, which corresponds in an optimal way to our solution possibilities? Is it possible to discover within our mathematical model and formulation already some qualitative or even quantitative results?

A perfect mechanical model, even a complicated one, will always be the simplest one possible, according to the well-known statement, that technology will be perfect if you cannot leave out anything. Especially for very complex systems we recommend to always start with a drastically simplified model for a better overview of the problems involved. Then, in a second step, establishing a large model will be easier. A good comprehension of the problems will always result in a better and faster development process.

1.3 Basic Concepts

1.3.1 Mass

We consider dynamics in the sense as discussed in Section 1.1. That means we do not refer to relativistic aspects whatsoever. The only deviation from the classical mass concept consists in the effects generated by rocket systems with their time-dependent masses. Focusing our future considerations mainly to technical artifacts, we usually know all relevant mass distributions and define:

- Masses are always positive, also in the time-dependent case, $m > 0$.
- Masses are
  - either constant with $\dot{m} = 0$,
  - or not constant with $\dot{m} \neq 0$, where $(\dot{m} = \frac{dm}{dt})$.
- Masses can be added and divided into parts.

Another more physically oriented definition of mass is given by Synge [66]. He states, that a mass is “a quantity of matter in a body, a measure of the reluctance of a body to change its velocity and a measure of the capacity of a body to attract another gravitationally”.

Modeling masses depends on the problem under consideration. We might have rigid or elastic masses and in dynamics also interactions with fluid masses. Theoretically, we always get an interdependence of the selected mass model and the results we can achieve with such a model. However for many practical cases, the experience of modeling tells us how to choose mass models. Nevertheless it makes sense to keep in mind these connections. In the following, we mainly consider systems with constant masses.

1.3.2 Euler’s Cut Principle and Forces

In mechanics, we are interested in the interaction of bodies with forces or torques. If we therefore separate two bodies by isolating them we must at the same time arrange forces along the cutting line, which in the original configuration keeps the two bodies together. Thus by establishing free body diagrams, we transform internal forces to external ones acting on both sides of the cutting line with the same magnitude but opposite sign. This ingenious cut principle, first established by EULER, was characterized by Szabo [67] in a very appropriate way: “EULER teaches us with the imagination of an artist to look in thought into the matter, where no eye and no experiment can enter. With this he has laid a foundation for the only genuine mechanics, namely continuum mechanics.” The cut principle gives us the opportunity to establish the equations of motion for any part of a system, if we choose the cutting lines correctly and add to the applied forces and torques also the reaction forces and torques as isolated by these cuts. We need in addition a sign convention, which we may choose arbitrarily, but we must stay with it.

The cut principle allows us, to separate masses and mass systems from their environment. To illustrate the difference of internal and external forces depending on the cutting line selection, we use a simple example (Fig. 1.1). Considering the cutting line 1 around the three masses, we see that all forces within that line are internal forces possessing no influence on the system 1. Selecting line 2, we come out with two external forces F_{12} and F_{32} and with two internal forces F_{13} and F_{31}. Finally, the line selection 3 generates only external forces, namely F_{21} and F_{31}.

The mechanical sciences are interested in the interaction of masses with forces. Dynamics as a part of mechanics is especially interested in those forces, which generate motion. Therefore, we define the concept of active and passive forces. Active forces can be moved in their direction of action, and from there they produce work and power. Passive forces cannot be moved with respect to their point of action. Active forces generate motion, passive forces prevent motion, they are the consequence of constraints. All other definitions of forces are subsets of this concept. Internal or external forces, applied or constraint forces, volume or surface forces, they all may be active or passive, depending on the specific system under consideration.

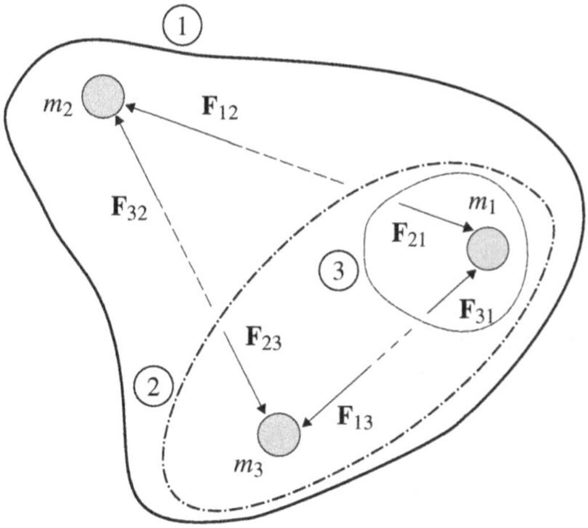

Fig. 1.1. EULER's cut principle and internal/external forces [16].

1.3.3 *Constraints*

If force interactions represent the very heart of mechanics, then constraints guarantee that the blood circulates through the vascular system. Constraints always possess a kinematic character. They are the mechanical controllers that tell systems where to go and where not to go. In mechanical engineering, there is no machine or mechanism that is not constrained. Constraints realize, at least kinematically, operational requirements and, applied correctly, guarantee the function of a mechanical system. Constraints can be bilateral or unilateral. In the first case, they represent ideal connections between two adjacent bodies, or between one body of the system and its environment, and reduce the degree of freedom of the system. In the second case, a connection may be open or closed, it may stick or slide, depending on the dynamics of the system under consideration. Some typical examples are depicted in Fig. 1.2 (pendulum), Fig. 1.3 (sledge), and in Fig. 1.4 (wheel).

Nearly all mechanical systems of practical relevancy are governed by a certain number of constraints, which depend on position, orientation, velocity, and time. Other forms of constraint do not exist, because it is clear from practical considerations that constraints describe only kinematic connection types. Constraints on an acceleration level are mathematically differentiated constraints and not in their original "physical" form. They are important in the theory of multibody systems.

On the basis of *position coordinates* $\mathbf{z} \in \mathbb{R}^\delta$, *velocity coordinates* $\dot{\mathbf{z}} \in \mathbb{R}^\delta$, *time* t, and *constraint functions* $\boldsymbol{\Phi} \in \mathbb{R}^m$ with $m < \delta$, we give some structure to these constraints [28, 68]. If the constraints depend explicitly on time, we call them *rheonomic*. A constraint that does not depend on time is named *scleronomic*.

Constraints that depend only on position or orientation but not on velocity are *holonomic constraints*:

- holonomic-scleronomic constraints

$$\boldsymbol{\Phi}(\mathbf{z}) = \mathbf{0}, \tag{1.1}$$

- holonomic-rheonomic constraints

$$\boldsymbol{\Phi}(\mathbf{z},t) = \mathbf{0}. \tag{1.2}$$

Every constraint might be differentiated with respect to time. Mathematically, these equations are called *hidden constraints*. They possess no physical meaning but have the character of invariants with the following linear structure:

$$\mathbf{0} = \dot{\boldsymbol{\Phi}}(\mathbf{z},\dot{\mathbf{z}},t) = \mathbf{W}(\mathbf{z},t)^T \dot{\mathbf{z}} + \hat{\mathbf{w}}(\mathbf{z},t), \tag{1.3}$$

$$\mathbf{0} = \ddot{\boldsymbol{\Phi}}(\mathbf{z},\dot{\mathbf{z}},\ddot{\mathbf{z}},t) = \mathbf{W}(\mathbf{z},t)^T \ddot{\mathbf{z}} + \bar{\mathbf{w}}(\mathbf{z},\dot{\mathbf{z}},t). \tag{1.4}$$

Using such hidden constraints we know that they can be integrated to position level. Then, the following conditions according to SCHWARZ [58] have to be satisfied:

$$\frac{\partial \mathbf{W}_{ik}^T}{\partial \mathbf{z}_j} = \frac{\partial^2 \Phi_i}{\partial \mathbf{z}_k \partial \mathbf{z}_j} = \frac{\partial^2 \Phi_i}{\partial \mathbf{z}_j \partial \mathbf{z}_k} = \frac{\partial \mathbf{W}_{ij}^T}{\partial \mathbf{z}_k} \quad \text{and} \quad \frac{\partial \mathbf{W}_{ik}^T}{\partial t} = \frac{\partial^2 \Phi_i}{\partial \mathbf{z}_k \partial t} = \frac{\partial^2 \Phi_i}{\partial t \partial \mathbf{z}_k} = \frac{\partial \hat{\mathbf{w}}_i}{\partial \mathbf{z}_k}. \tag{1.5}$$

Conversely, for the case $\mathbf{z} \in \mathrm{IR}^{\delta}$, these conditions can also be used to decide whether an integration to position level is possible [58]. However, in applications, we often find constraints that are exclusively linear in some velocities with the property, that they cannot be integrated to position level. These are *nonholonomic constraints*. They can be expressed in the form:

- nonholonomic-scleronomic constraints

$$\boldsymbol{\Phi}(\mathbf{z},\dot{\mathbf{z}}) = \mathbf{W}(\mathbf{z})^T \dot{\mathbf{z}} + \hat{\mathbf{w}}(\mathbf{z}) = \mathbf{0}, \tag{1.6}$$

- nonholonomic-rheonomic constraints

$$\boldsymbol{\Phi}(\mathbf{z},\dot{\mathbf{z}},t) = \mathbf{W}(\mathbf{z},t)^T \dot{\mathbf{z}} + \hat{\mathbf{w}}(\mathbf{z},t) = \mathbf{0}. \tag{1.7}$$

Constraints of real physical significance, which include nonlinear relations in the velocities, are not known.

Example 1.1 (holonomic-scleronomic constraints of the pendulum). The particle mass of the string pendulum in Fig. 1.2 moves on a circular trajectory with the radius R. This trajectory is described by its Cartesian position $\mathbf{z} := \mathbf{r}$ with respect to a reference O. It moves in a gravity field with gravitational acceleration $\mathbf{g}$. Cutting the string (Section 1.1), for example, leads to two forces acting on the pendulum mass; the gravity force $m\mathbf{g}$ and a constraint force $\mathbf{F}^z$, which forces the mass into

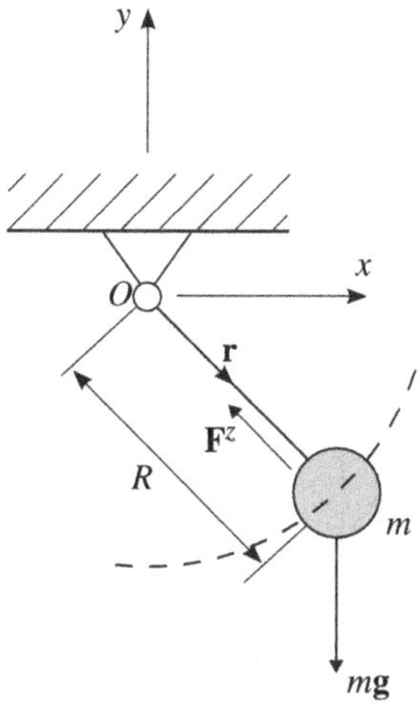

Fig. 1.2. Holonomic constraints of the string pendulum.

the circular trajectory and, being a passive force, does not contribute to motion. The relevant equations are:

$$m\ddot{\mathbf{r}} = m\mathbf{g} + \mathbf{F}^z,$$
$$\Phi(\mathbf{r}) = \mathbf{r}^T\mathbf{r} - R^2 = 0.$$

In the following sections we present some more complete explanations.

Example 1.2 (holonomic-scleronomic constraints for the sledge). To a large extent the motion of a sledge is governed by ground contours, which are thus responsible for the resulting constraint $\Phi(\mathbf{r}) = \mathbf{0}$ (Fig. 1.3). The position of the sledge is described in Cartesian coordinates with $\mathbf{z} := \mathbf{r}$. The interaction with the ground is twofold. For the tangential direction, we have to consider frictional effects, and for the normal direction, we must consider the normal constraint force as a passive force interacting with the ground and keeping the sledge on the ground contour. Assuming sliding of the sledge governed by COULOMB's law of friction yields the sliding friction force $\mathbf{W}_T(\mathbf{r})\mu\lambda_N$. The magnitude $\mathbf{W}_T(\mathbf{r})$ represents the relevant normalized tangential direction, $\mu \geq 0$ is the coefficient of sliding friction, and λ_N is a force parameter resulting from the constraint in the relevant normalized normal direction $\mathbf{W}_N(\mathbf{r})$. We get

$$m\ddot{\mathbf{r}} = m\mathbf{g} + \mathbf{W}_T(\mathbf{r})\mu\lambda_N + \mathbf{W}_N(\mathbf{r})\lambda_N,$$
$$\Phi(\mathbf{r}) = 0.$$

We see that because of friction the parameter λ_N is involved in the free motion of the sledge in an active way. Therefore, $(\mu\lambda_N)$ is not a constraint force in the classical sense. It participates in the energy budget of the system and is an active force. For

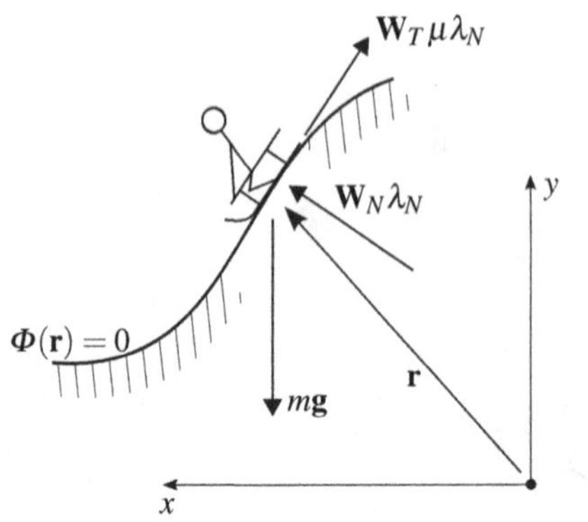

Fig. 1.3. Holonomic constraints of the sledge.

$\mu = 0$, we are left only with the constraint force λ_N forcing the sledge onto its trajectory (Section 1.7).

Example 1.3 (Nonholonomic-scleronomic constraints of a wheel). We consider a disc with radius R rolling on a plain (Fig. 1.4). In this case, we describe the disc using generalized coordinates $\mathbf{z} := (x, y, \alpha, \beta)^T$ and not Cartesian ones. Considering the rolling condition we get

$$\mathrm{d}s = R\,\mathrm{d}\beta,$$

and from this

$$\mathrm{d}x = \mathrm{d}s\ \cos\alpha = R\ \cos\alpha\ \mathrm{d}\beta,$$
$$\mathrm{d}y = \mathrm{d}s\ \sin\alpha = R\ \sin\alpha\ \mathrm{d}\beta.$$

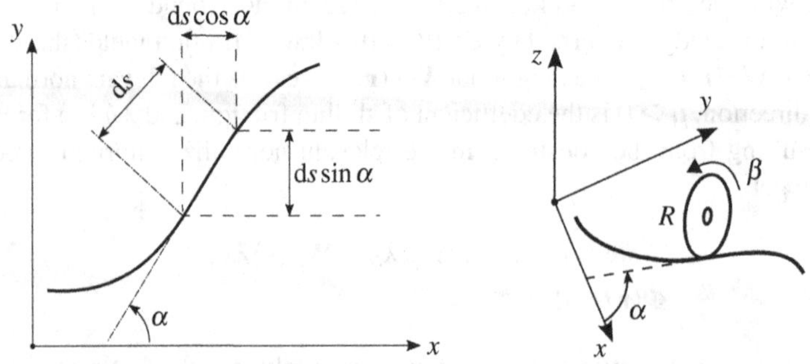

Fig. 1.4. Nonholonomic constraints of a rolling disc.

Division by dt results in the following nonholonomic-scleronomic constraints:

$$\boldsymbol{\Phi}(\mathbf{z},\dot{\mathbf{z}}) = \underbrace{\begin{pmatrix} 1 & 0 & 0 & -R\cos\alpha \\ 0 & 1 & 0 & -R\sin\alpha \end{pmatrix}}_{\mathbf{W}^T} \underbrace{\begin{pmatrix} \dot{x} \\ \dot{y} \\ \dot{\alpha} \\ \dot{\beta} \end{pmatrix}}_{\dot{\mathbf{z}}} = \mathbf{0}.$$

For example,

$$R\sin\alpha = \frac{\partial \mathbf{W}_{14}^T}{\partial \alpha} \neq \frac{\partial \mathbf{W}_{13}^T}{\partial \beta} = 0.$$

The integration condition is not satisfied and the constraint is not integrable to position level. If $\alpha \equiv \alpha_0$ is constant, we can integrate the constraints resulting in a set of holonomic-scleronomic equations

$$\begin{aligned} (x - x_0) - R\cos\alpha_0\ (\beta - \beta_0) &= 0, \\ (y - y_0) - R\sin\alpha_0\ (\beta - \beta_0) &= 0 \end{aligned}$$

with integration constants x_0, y_0, and β_0. The restriction $\alpha \equiv \alpha_0$ can be interpreted as rolling on a straight trajectory, where rolling forward and backward always results in the same starting position.

Constraints form constraint surfaces and generate constraint forces, which realize a certain motion structure as drawn up by the designer of the system. Examining the examples and with some generalization, we conclude that these constraint forces are always perpendicular to the constraint surfaces. This property will be the basis for the principle of d'ALEMBERT, which is dealt with later (Section 1.7). Kinetically, motion can only take place on these constraint surfaces, and never perpendicular to them.

1.3.4 Virtual Displacements

The concept of virtual displacement can to a certain extent be compared to the concept of *variation* in the *calculus of variations* [14]. The basic idea can be visualized by considering a snap-shot of the generalized configuration $\mathbf{z}$ of a dynamic system at a fixed time t. Then, a virtual displacement of a system is an arbitrary, imaginary, and small change $\delta\mathbf{z}$ compatible with the existing constraints [10, 59]. Using this thought process, we have to bear in mind that virtual velocity changes need not necessarily be small, but compatibility with constraints must always be satisfied. We come back to this point later. Conversely, a *real displacement* of the system under consideration takes places within the time interval dt, where all states, forces, and constraints may be changing dependent on time.

Example 1.4 (Spherical pendulum). The point mass of a mathematical pendulum moves on the surface of a sphere with radius R (Fig. 1.5).

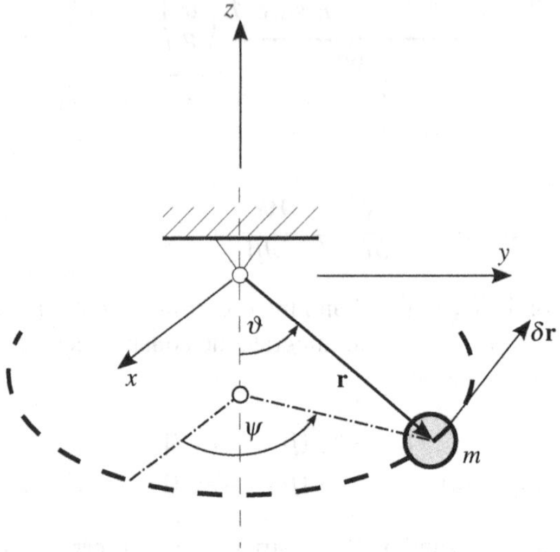

Fig. 1.5. Virtual displacement of a spherical pendulum, ($\| \mathbf{r} \| = R$).

We use the Cartesian position $\mathbf{z} := \mathbf{r} = (x,y,z)^T$ for a description of the pendulum mass by watching also the constraint $\phi(\mathbf{r}) = \mathbf{r}^T\mathbf{r} - R^2 = x^2 + y^2 + z^2 - R^2 \stackrel{!}{=} 0$. This means that a small virtual displacement of the mass necessarily takes place "within this spherical surface." Up to terms of first order, we have

$$0 \stackrel{!}{=} \phi(\mathbf{r} + \delta\mathbf{r}) \doteq \underbrace{\phi(\mathbf{r})}_{=0} + \underbrace{\frac{\partial\phi}{\partial\mathbf{r}}\delta\mathbf{r}}_{=:\delta\phi} = 2\mathbf{r}^T\delta\mathbf{r} = 2(x\delta x + y\delta y + z\delta z).$$

This is a fundamental statement telling us with the term $\mathbf{r}^T\delta\mathbf{r} = 0$ that a virtual displacement $\delta\mathbf{r}$ compatible with the constraints always has to be perpendicular to the normal space of the constraint surface, that is $\delta\mathbf{r} \perp \mathbf{r}$. In the example of the spherical pendulum, this "constraint surface" is the sphere with radius R.

1.4 Kinematics

1.4.1 Coordinate Systems and Coordinates

We describe the motion of bodies by coordinate systems (Fig. 1.6). A *Cartesian coordinate system* is a set of orthogonal *unit vectors* $\mathbf{e}_x$, $\mathbf{e}_y$, $\mathbf{e}_z$, which form a basis for

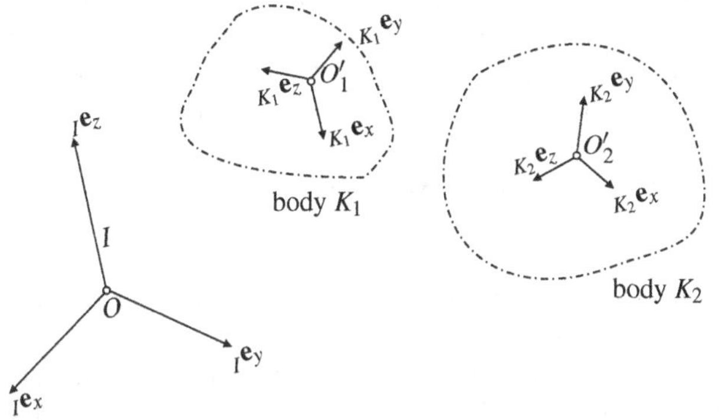

Fig. 1.6. Translation and rotation of coordinates, I for "inertial".

all vectors in an inertial space. This basis of the vector space includes some origin O, which we may think of as being connected to some body (for example, some point on the earth's surface). The definition of an origin allows us to measure geometric magnitudes and from this also dynamic processes. Depending on the state of motion of a coordinate system $(O, \mathbf{e}_x, \mathbf{e}_y, \mathbf{e}_z)$, we call it inertial or noninertial (body-fixed).

The coordinate system $(O, \mathbf{e}_x, \mathbf{e}_y, \mathbf{e}_z)$ possesses the property "inertial", if the basis vectors $(\mathbf{e}_x, \mathbf{e}_y, \mathbf{e}_z)$ do not change with time, which means, that such a coordinate system might only move with constant velocity with respect to the postulated non-moving space. This is a question of definition. For technical dynamics, it is usually sufficient to connect the earth or some building with an inertial coordinate system, for problems of space dynamics the sun might be a more suitable system. Often the coordinate system can be assumed to be even space-fixed. We use the index I.

If we connect a coordinate system with a body to define a “body-fixed” coordinate system, we can choose to select any point for the origin. The center of mass or another convenient point may be suitable for our evaluations, for example a joint in the case of robots. We use the index K. We may describe several coordinate systems for one body. A basic property of a rigid body is the constant distance between two material points. Rigid bodies have six degrees of freedom, three of which are translational and three rotational. Therefore, the three positions (x,y,z) and the three orientations (α, β, γ), given for example as Cardan angles, define the position and orientation of a rigid body with respect to any coordinate system, inertial or body-fixed. The position and the orientation of a body K_1 can be described with respect to the inertial system I or with respect to the body-fixed system of body K_2 by the six magnitudes (x,y,z) and (α, β, γ), where (x,y,z) are the coordinates of the center of mass of the body K_1, written in the bases I or K_2, and where the Cardan angles (α, β, γ) give the orientation between the coordinate system of K_1 and those of I or K_2, written correspondingly in the I- or K_2-bases (see Fig. 1.6).

Returning to Fig. 1.6, we see that the motion of a body may be described by a relative *translation* of the origin O' and by a relative *rotation* of the body-fixed coordinate system K with respect to the reference coordinate system $R = I$ (inertial system) with origin O. The coordinates of O' defined in R are ${}_Rx$, ${}_Ry$, ${}_Rz$. The relative rotation of the body-fixed unit vectors ${}_K\mathbf{e}_x$, ${}_K\mathbf{e}_y$, ${}_K\mathbf{e}_z$ are formally described by the column vectors of the rotation matrix $\mathbf{A}_{RK} \in \mathrm{I\!R}^{3,3}$. Rotation matrices are orthogonal and possess the properties (the indices 'KR' are defined after (1.23)):

$$\mathbf{A}_{KR} = \mathbf{A}_{RK}^{-1} = \mathbf{A}_{RK}^T, \tag{1.8}$$

$$\det \mathbf{A}_{RK} = 1. \tag{1.9}$$

The rotation is described by nine "angular magnitudes," which depend on each other. Equation (1.8) defines six constraints according to Section 1.3.3, and (1.9) defines an additional sign condition. Altogether we are left with three translational *degrees of freedom* and also with three ($3 = 9 - 6$) rotational degrees of freedom. Examples of such minimal angle parametrizations are the EULER *angles* (Fig. 1.7)

$$\mathbf{A}_{RK} = \mathbf{A}_z^T(\psi)\mathbf{A}_x^T(\vartheta)\mathbf{A}_z^T(\varphi)$$
$$= \begin{pmatrix} \cos\varphi\cos\psi - \cos\vartheta\sin\psi\sin\varphi & -\sin\varphi\cos\psi - \cos\vartheta\sin\psi\cos\varphi & \sin\vartheta\sin\psi \\ \cos\varphi\sin\psi + \cos\vartheta\cos\psi\sin\varphi & -\sin\varphi\sin\psi + \cos\vartheta\cos\psi\cos\varphi & -\sin\vartheta\cos\psi \\ \sin\vartheta\sin\varphi & \sin\vartheta\cos\varphi & \cos\vartheta \end{pmatrix}, \tag{1.10}$$

with the *elementary rotations*

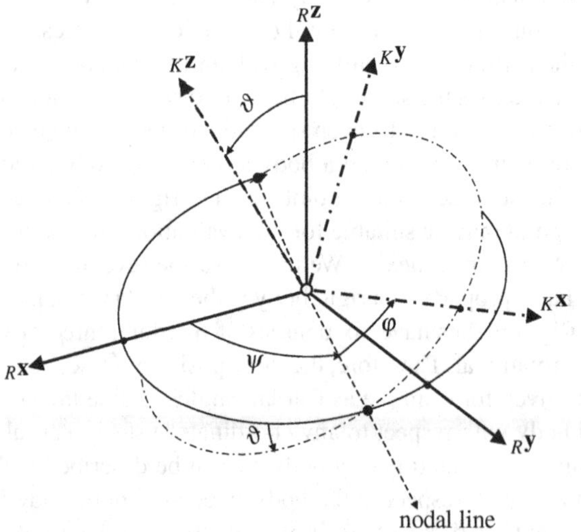

Fig. 1.7. EULER angles with azimuth- or precession angle ψ, elevation- or nutation angle ϑ, and rotation angle φ.

$$\mathbf{A}_z(\psi) = \begin{pmatrix} \cos\psi & \sin\psi & 0 \\ -\sin\psi & \cos\psi & 0 \\ 0 & 0 & 1 \end{pmatrix}, \tag{1.11}$$

$$\mathbf{A}_x(\theta) = \begin{pmatrix} 1 & 0 & 0 \\ 0 & \cos\theta & \sin\theta \\ 0 & -\sin\theta & \cos\theta \end{pmatrix}, \tag{1.12}$$

$$\mathbf{A}_z(\varphi) = \begin{pmatrix} \cos\varphi & \sin\varphi & 0 \\ -\sin\varphi & \cos\varphi & 0 \\ 0 & 0 & 1 \end{pmatrix} \tag{1.13}$$

and the *Cardan angles* (Fig. 1.8)

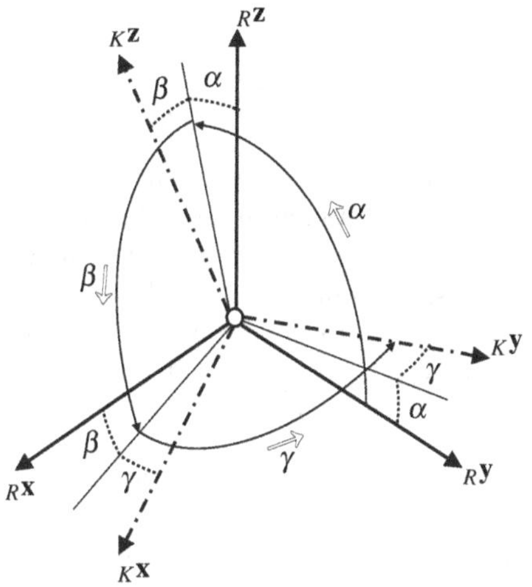

Fig. 1.8. Cardan angles.

$$\begin{aligned} \mathbf{A}_{RK} &= \mathbf{A}_x^T(\alpha)\mathbf{A}_y^T(\beta)\mathbf{A}_z^T(\gamma) \\ &= \begin{pmatrix} \cos\beta\cos\gamma & -\cos\beta\sin\gamma & \sin\beta \\ \cos\alpha\sin\gamma+\sin\alpha\sin\beta\cos\gamma & \cos\alpha\cos\gamma-\sin\alpha\sin\beta\sin\gamma & -\sin\alpha\cos\beta \\ \sin\alpha\sin\gamma-\cos\alpha\sin\beta\cos\gamma & \sin\alpha\cos\gamma+\cos\alpha\sin\beta\sin\gamma & \cos\alpha\cos\beta \end{pmatrix} \end{aligned} \tag{1.14}$$

with the elementary rotations

$$\mathbf{A}_x(\alpha) = \begin{pmatrix} 1 & 0 & 0 \\ 0 & \cos\alpha & \sin\alpha \\ 0 & -\sin\alpha & \cos\alpha \end{pmatrix}, \tag{1.15}$$

$$\mathbf{A}_y(\beta) = \begin{pmatrix} \cos\beta & 0 & -\sin\beta \\ 0 & 1 & 0 \\ \sin\beta & 0 & \cos\beta \end{pmatrix}, \tag{1.16}$$

$$\mathbf{A}_z(\gamma) = \begin{pmatrix} \cos\gamma & \sin\gamma & 0 \\ -\sin\gamma & \cos\gamma & 0 \\ 0 & 0 & 1 \end{pmatrix}. \tag{1.17}$$

We may also apply (1.8) and (1.9) as they use a kind of natural parametrization [30]. We collect the descriptive parameters with respect to some coordinate system in a vector of *generalized coordinates*. Using for example Cardan angles, we come out with

$$\mathbf{z} := ({}_Rx, {}_Ry, {}_Rz, \alpha, \beta, \gamma)^T \in \mathrm{I\!R}^6. \tag{1.18}$$

Considering a rigid body, we know, that any distance between two points of the rigid body is constant enabling a description of any relative motion on it by one body-fixed coordinate system together with appropriate relative coordinates. For more complex systems of masses, for example multibody systems, we must also introduce more complex systems of descriptive coordinates. The position and orientation of many coordinate systems $K_1, \ldots, K_n$ are then defined in a *configuration space* of dimension (δn):

$$\mathbf{z} := (\mathbf{z}_1^T, \ldots, \mathbf{z}_n^T)^T \in \mathrm{I\!R}^{\delta n}. \tag{1.19}$$

If we are not able to find a set of minimal coordinates, we have to additionally consider a corresponding set of constraints of the form (Section 1.3.3)

$$\phi_i(\mathbf{z}) = 0 \text{ for } i \in \{1, \ldots, m\} \text{ and } m < \delta n. \tag{1.20}$$

In many cases of practical relevancy, we may find a complete set $\mathbf{q}$ of minimal coordinates with the number $f := \delta n - m$ of degrees of freedom. However, this will be not possible in every case, and we should also bear in mind, that there are no precise rules for the detection of minimal coordinates. The corresponding mathematical formulae do not necessarily replace a physical search. Expressing $\mathbf{z}$ by the minimal coordinates $\mathbf{q}$ we obtain

$$\phi_i(\mathbf{z}(\mathbf{q})) = 0 \quad \forall \mathbf{q} \in \mathrm{I\!R}^f \tag{1.21}$$

for $i \in \{1, \ldots, m\}$. Summarizing all these coordinates we write

- $\mathbf{r}$ Cartesian position of some point,
- $\mathbf{z}$ generalized coordinates,
- $\mathbf{q}$ minimal coordinates.

The derivations $\dot{\mathbf{r}}$, $\dot{\mathbf{z}}$, $\dot{\mathbf{q}}$ with respect to time are called *velocity coordinates*.

1.4.2 Transformation of Coordinates

The kinematic relations always form a basis for generating the equations of motion. For large dynamic systems, we get various sets of coordinates, which require transformations from one to the other and from some inertial basis to some body-fixed coordinate systems. Typically, the fundamental kinetic laws apply with respect to an inertial, space-fixed basis, whereas bodies are better described in a body-fixed frame. As a consequence, we need transformations between all existing coordinates. Performing the corresponding processes, we should keep in mind, that a vector sticks to a vector in spite of the fact that it will be described by different coordinates in different coordinate systems.

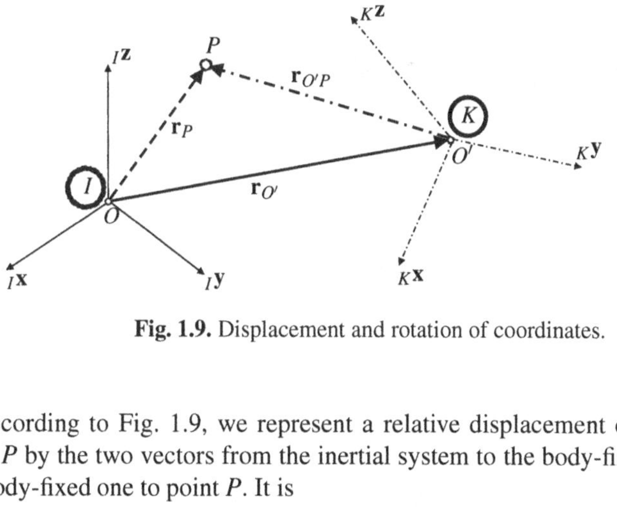

Fig. 1.9. Displacement and rotation of coordinates.

According to Fig. 1.9, we represent a relative displacement of the body-fixed point P by the two vectors from the inertial system to the body-fixed one and from the body-fixed one to point P. It is

$$_K\mathbf{r}_P = {_K\mathbf{r}_{O'}} + {_K\mathbf{r}_{O'P}}. \tag{1.22}$$

In the following text, we omit the indices O or I, if they refer to some inertial coordinates.

According to Section 1.4.1, the rotation of coordinate systems is described by a rotation matrix, which also defines the transformation matrix. It transforms the vector $_K\mathbf{r}_P$ of the K-system into a vector $_I\mathbf{r}_P$ of the I-system [10, 40]:

$$_I\mathbf{r}_P = \mathbf{A}_{IK}\,{_K\mathbf{r}_P}. \tag{1.23}$$

The double index IK means the following: matrix $\mathbf{A}$ transforms vector $\mathbf{r}_P$, given in the K-system, into the I-system. All similar double indices should be interpreted in the same way.

1.4.3 Relative Kinematics

From the discussion above it is clear, that we need all absolute and relative velocities and accelerations of the bodies under consideration and of their coordinates. Therefore, the main goal in establishing *relative kinematics* (Fig. 1.10) consists in evaluating these magnitudes.

This is formal work provided we have established before a precise geometric description of the whole system, including a convenient choice of coordinates and a best possible set of constraints based on a good choice of interconnections. NEWTON's "mass times acceleration is equal to force" is simple, but the accelerations can be awfully complex. Let us start with (1.22), (1.23), and Fig. 1.10. From there we write

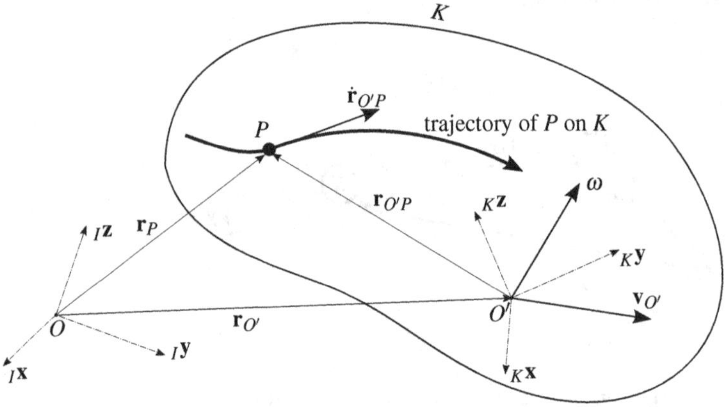

Fig. 1.10. Relative kinematics of a body.

$$_I\mathbf{r}_P = {_I\mathbf{r}_{O'}} + \mathbf{A}_{IK}\, {_K\mathbf{r}_{O'P}}\,. \tag{1.24}$$

Deriving this formally results in

$$_I\dot{\mathbf{r}}_P = {_I\dot{\mathbf{r}}_{O'}} + \dot{\mathbf{A}}_{IK}\, {_K\mathbf{r}_{O'P}} + \mathbf{A}_{IK}\, {_K\dot{\mathbf{r}}_{O'P}}\,, \tag{1.25}$$

$$_I\ddot{\mathbf{r}}_P = {_I\ddot{\mathbf{r}}_{O'}} + \ddot{\mathbf{A}}_{IK}\, {_K\mathbf{r}_{O'P}} + 2\dot{\mathbf{A}}_{IK}\, {_K\dot{\mathbf{r}}_{O'P}} + \mathbf{A}_{IK}\, {_K\ddot{\mathbf{r}}_{O'P}}\,. \tag{1.26}$$

The vector $_I\mathbf{v}_P = {_I\dot{\mathbf{r}}_P}$ represents the *absolute velocity* of point P in the I-system and the vector $_K\dot{\mathbf{r}}_{O'P}$ its *relative velocity* as seen from O' in the K-system. The *translational velocity* of point O' is an absolute velocity $_I\mathbf{v}_{O'} = {_I\dot{\mathbf{r}}_{O'}}$ represented with respect to the I-system.

In a similar sense, the vector $_I\mathbf{a}_P = {_I\ddot{\mathbf{r}}_P}$ is the *absolute acceleration* of point P in the I-system and $_K\ddot{\mathbf{r}}_{O'P}$ its *relative acceleration* as seen from O' in the K-system. Also $_I\mathbf{a}_{O'} = {_I\ddot{\mathbf{r}}_{O'}}$ is the absolute acceleration of point O' in the I-system. Going to the body-fixed frame, we have to multiply these relations by $\mathbf{A}_{KI}$ and get

$$
\begin{aligned}
{}_K\mathbf{v}_P &= {}_K\mathbf{v}_{O'} + \mathbf{A}_{IK}^T\dot{\mathbf{A}}_{IK}\,{}_K\mathbf{r}_{O'P} + {}_K\dot{\mathbf{r}}_{O'P}\,, &(1.27)\\
{}_K\mathbf{a}_P &= {}_K\mathbf{a}_{O'} + \mathbf{A}_{IK}^T\ddot{\mathbf{A}}_{IK}\,{}_K\mathbf{r}_{O'P} + 2\mathbf{A}_{IK}^T\dot{\mathbf{A}}_{IK}\,{}_K\dot{\mathbf{r}}_{O'P} + {}_K\ddot{\mathbf{r}}_{O'P}\,. &(1.28)
\end{aligned}
$$

Formal derivation of the identity matrix $\mathbf{E} = \mathbf{A}_{IK}^T\mathbf{A}_{IK}$ results in

$$
\mathbf{0} = \dot{\mathbf{E}} = \mathbf{A}_{IK}^T\dot{\mathbf{A}}_{IK} + \dot{\mathbf{A}}_{IK}^T\mathbf{A}_{IK} \tag{1.29}
$$

demonstrating the skew-symmetric properties of

$$
{}_K\tilde{\omega} := \mathbf{A}_{IK}^T\dot{\mathbf{A}}_{IK} = -\left(\mathbf{A}_{IK}^T\dot{\mathbf{A}}_{IK}\right)^T = -{}_K\tilde{\omega}^T\,. \tag{1.30}
$$

It means that ${}_K\tilde{\omega}$ can be represented by three independent components only. Combining this in a vector ${}_K\omega$ yields

$$
{}_K\tilde{\omega}\,{}_K\mathbf{r}_{O'P} = {}_K\omega \times {}_K\mathbf{r}_{O'P} \text{ and } {}_K\tilde{\omega} = \begin{pmatrix} 0 & -{}_K\omega_3 & {}_K\omega_2 \\ {}_K\omega_3 & 0 & -{}_K\omega_1 \\ -{}_K\omega_2 & {}_K\omega_1 & 0 \end{pmatrix}. \tag{1.31}
$$

Then, we get from (1.27)

$$
{}_K\mathbf{v}_P = {}_K\mathbf{v}_{O'} + {}_K\omega \times {}_K\mathbf{r}_{O'P} + {}_K\dot{\mathbf{r}}_{O'P}\,. \tag{1.32}
$$

We come back to Fig. 1.10 with the moving body K and understand this figure as a moving base with a man walking on it, for example some kind of merry-go-round. The man is represented by point P, moving along some trajectory. The absolute velocity ${}_K\mathbf{v}_P$ of point P has two parts, first a relative velocity of P with respect to O' and second an *applied velocity* resulting from the body's motion with respect to the inertial system, which writes

$$
{}_K\mathbf{v}_{O'} + {}_K\omega \times {}_K\mathbf{r}_{O'P}\,. \tag{1.33}
$$

The rotational term ${}_K\omega \times {}_K\mathbf{r}_{O'P}$ is generated by the *angular velocity* ${}_K\omega$ of the body K written in a body-fixed frame with the distance ${}_K\mathbf{r}_{O'P}$ of the point P from O'. By not considering the translational velocity of point O' in (1.32), we get as a result the absolute rate of change of ${}_K\mathbf{r}_{O'P}$ with time:

$$
{}_K\mathbf{v}_P = {}_K\dot{\mathbf{r}}_{O'P} + {}_K\omega \times {}_K\mathbf{r}_{O'P}\,. \tag{1.34}
$$

This equation is one of the fundamental relations of relative kinematics. It applies to every vector in and on a moving system like the body K, not only to the special vector ${}_K\mathbf{r}_{O'P}$. It is called *Euler's theorem* or the *Coriolis equation*. In words:

The absolute rate of change of a vector with respect to time is equal to the sum of its applied rate of change and of its relative rate of change [41].

The determination of the angular velocities may be performed formally using the rotation matrices and their derivations, but there is a nice and more elegant way to derive these angular velocities by considering the rotation sequences as depicted in Figs. 1.7 and 1.8.

Starting with the EULER angles, we notice from Fig. 1.7 the following sequence of rotations: In a first step we rotate with the angle ψ about the z-axis and come to an end-position in the form of a nodal line. From there we tilt with ϑ about this nodal line, and then finally we rotate with φ about the final z-axis. To achieve a representation in inertial coordinates, we must transform back the second and the third rotations into a global frame. We get

$$
\begin{aligned}
{}_I\omega &= \mathbf{e}_z\dot{\psi} + \mathbf{A}_z^T(\psi)\mathbf{e}_x\dot{\vartheta} + \mathbf{A}_z^T(\psi)\mathbf{A}_x^T(\vartheta)\mathbf{e}_z\dot{\varphi} \\
&= \begin{pmatrix} \sin\psi\sin\vartheta & \cos\psi & 0 \\ -\cos\psi\sin\vartheta & \sin\psi & 0 \\ \cos\vartheta & 0 & 1 \end{pmatrix} \begin{pmatrix} \dot{\varphi} \\ \dot{\vartheta} \\ \dot{\psi} \end{pmatrix}.
\end{aligned}
$$

In a similar way, we do that for the Cardan angles. We rotate first with α about the x-axis, second with β about the just-generated y-axis, and third with γ about the new z-axis. Then, we come out with

$$
\begin{aligned}
{}_I\omega &= \mathbf{e}_x\dot{\alpha} + \mathbf{A}_x^T(\alpha)\mathbf{e}_y\dot{\beta} + \mathbf{A}_x^T(\alpha)\mathbf{A}_y^T(\beta)\mathbf{e}_z\dot{\gamma} \\
&= \begin{pmatrix} 1 & 0 & \sin\beta \\ 0 & \cos\alpha & -\sin\alpha\cos\beta \\ 0 & \sin\alpha & \cos\alpha\cos\beta \end{pmatrix} \begin{pmatrix} \dot{\alpha} \\ \dot{\beta} \\ \dot{\gamma} \end{pmatrix}.
\end{aligned}
$$

Some relationship for angular velocities dependent on a set of rotation parameters $\bar{\varphi}$, for example $\bar{\varphi} = (\varphi, \vartheta, \psi)^T$ or $\bar{\varphi} = (\alpha, \beta, \gamma)^T$, can always be established in the form

$$
{}_I\omega = \mathbf{Y}_I^{-1}\dot{\bar{\varphi}},
$$

but the inverse of this matrix $\mathbf{Y}_I^{-1}$ does not exist for all possible magnitudes of the relevant angles. For our two sets above, we have

$$
\mathbf{Y}_I = \frac{1}{\sin\vartheta}\begin{pmatrix} \sin\psi & -\cos\psi & 0 \\ \cos\psi\sin\vartheta & \sin\psi\sin\vartheta & 0 \\ -\sin\psi\cos\vartheta & \cos\psi\cos\vartheta & \sin\vartheta \end{pmatrix} \qquad \text{(EULER angles)},
$$

$$
\mathbf{Y}_I = \frac{1}{\cos\beta}\begin{pmatrix} \cos\beta & \sin\beta\sin\alpha & -\sin\beta\cos\alpha \\ 0 & \cos\beta\cos\alpha & \cos\beta\sin\alpha \\ 0 & -\sin\alpha & \cos\alpha \end{pmatrix} \qquad \text{(Cardan angles)}.
$$

These expressions possess *singularities*, which are typical for angular systems containing a minimal set of parameters. The singularity with respect to the Cardan angles is also called the "gimbal lock". The singularities for the EULER and Cardan angles are at $\vartheta = 0$ and $\beta = \frac{\pi}{2}$, respectively, which are in both cases special angles of the second elementary rotation. This can be generalized. If we design the second elementary rotation in such a way, that the first and third elementary rotation take place about the same spatial axis, then the complete rotation will come out with redundancy including a singularity. The singularity can be avoided by choosing another set of minimal parameters. It can be completely avoided by applying not a minimal set but a nonminimal set of rotation parameters, for example EULER parameters or quaternions [30, 5, 63].

Proceeding from relative velocities to relative accelerations, we consider (1.28) and the point P moving on K in Fig. 1.10. By formal derivation, we get

$$ {}_K\dot{\tilde{\omega}} = \frac{\mathrm{d}}{\mathrm{d}t}\left(\mathbf{A}_{IK}^T\dot{\mathbf{A}}_{IK}\right) = \dot{\mathbf{A}}_{IK}^T\underbrace{\mathbf{A}_{IK}\mathbf{A}_{IK}^T}_{\mathbf{E}}\dot{\mathbf{A}}_{IK} + \mathbf{A}_{IK}^T\ddot{\mathbf{A}}_{IK} = -{}_K\tilde{\omega}\,{}_K\tilde{\omega} + \mathbf{A}_{IK}^T\ddot{\mathbf{A}}_{IK} \tag{1.35} $$

and in combination with (1.28) the result

$$ {}_K\mathbf{a}_P = {}_K\mathbf{a}_{O'} + {}_K\dot{\omega}\times{}_K\mathbf{r}_{O'P} + {}_K\omega\times({}_K\omega\times{}_K\mathbf{r}_{O'P}) + 2\,{}_K\omega\times{}_K\dot{\mathbf{r}}_{O'P} + {}_K\ddot{\mathbf{r}}_{O'P}\,. \tag{1.36} $$

We may derive this equation also by applying EULER's theorem (1.32) concerning the absolute velocity

$$ \begin{aligned} {}_K\mathbf{a}_P &= {}_K\omega\times{}_K\mathbf{v}_P + {}_K\dot{\mathbf{v}}_P \\ &= {}_K\omega\times({}_K\mathbf{v}_{O'} + {}_K\omega\times{}_K\mathbf{r}_{O'P} + {}_K\dot{\mathbf{r}}_{O'P}) + \frac{\mathrm{d}}{\mathrm{d}t}({}_K\mathbf{v}_{O'} + {}_K\omega\times{}_K\mathbf{r}_{O'P} + {}_K\dot{\mathbf{r}}_{O'P}) \end{aligned} \tag{1.37} $$

together with some organization of the terms.

The absolute acceleration possesses three components:

- applied acceleration

$$ {}_K\mathbf{a}_{O'} + {}_K\dot{\omega}\times{}_K\mathbf{r}_{O'P} + {}_K\omega\times({}_K\omega\times{}_K\mathbf{r}_{O'P}) \tag{1.38} $$

  with the absolute acceleration of the moving body (reference) ${}_K\mathbf{a}_{O'}$, the rotational acceleration ${}_K\dot{\omega}\times{}_K\mathbf{r}_{O'P}$, and the centripedal acceleration ${}_K\omega\times({}_K\omega\times{}_K\mathbf{r}_{O'P})$,

- Coriolis acceleration

$$ 2\,{}_K\omega\times{}_K\dot{\mathbf{r}}_{O'P}\,, \tag{1.39} $$

- relative acceleration

$$ {}_K\ddot{\mathbf{r}}_{O'P}\,. \tag{1.40} $$

Example 1.5 (Rotating disc with a radially guided ball). A disc K turns with constant and positive angular velocity (${}_K\omega = \begin{pmatrix}0 & 0 & \omega\end{pmatrix}^T$,and $\omega > 0$) about a fixed center $O' = O$. A ball P moves with constant and positive velocity ($v > 0$) in a tube fixed on the disc, which means ${}_K\mathbf{r}_{O'P} = \begin{pmatrix}vt & 0 & 0\end{pmatrix}^T$. We look for the absolute

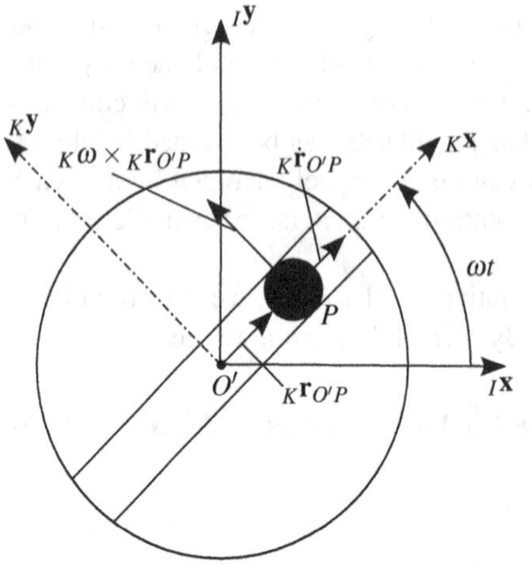

Fig. 1.11. Rotating disc with a radially guided ball.

velocity and the absolute acceleration of the ball, starting with the time $t = 0$. According to the relations above, we get for the absolute velocity

$$_K\mathbf{v}_P = {_K\omega} \times {_K\mathbf{r}_{O'P}} + {_K\dot{\mathbf{r}}_{O'P}} = \begin{pmatrix} v & \omega v t & 0 \end{pmatrix}^T$$

where the two terms are the relative velocity v of the ball in radial direction and its applied velocity ωvt in circumferential direction. The absolute acceleration writes

$$_K\mathbf{a}_P = {_K\omega} \times ({_K\omega} \times {_K\mathbf{r}_{O'P}}) + 2\,{_K\omega} \times {_K\dot{\mathbf{r}}_{O'P}} = \begin{pmatrix} -\omega^2 vt & 2\omega v & 0 \end{pmatrix}^T .$$

We recognize the centripedal acceleration $-\omega^2 vt$ in the negative radial direction and the Coriolis acceleration $2\omega v$ in circumferential direction, well known from the devil wheel of modern fairs. One part of the Coriolis acceleration comes from the derivation ${_K\dot{\mathbf{v}}_P} = \begin{pmatrix} 0 & \omega v & 0 \end{pmatrix}^T$, a second part comes from the angular motion ${_K\omega} \times {_K\mathbf{v}_P} = \begin{pmatrix} -\omega^2 vt & \omega v & 0 \end{pmatrix}^T$, which gives in addition the centripedal acceleration in negative radial direction.

We perform an experiment of thought. Sitting for example on this ball not knowing the rotation of the disc, then the ball is exposed only to its relative acceleration with respect to the disc. Considering now NEWTON's law, we must express the absolute relative acceleration, "felt" by the ball, by all its components as discussed before, namely

$$_K\mathbf{F}_a = m\,{_K\mathbf{a}_P} = m\,{_K\omega} \times ({_K\omega} \times {_K\mathbf{r}_{O'P}}) + m2\,{_K\omega} \times {_K\dot{\mathbf{r}}_{O'P}} + m\,{_K\ddot{\mathbf{r}}_{O'P}} .$$

Therefore, the ball "feels" the centrifugal force in radial direction and the Coriolis force in negative circumferential direction:

$$m_K \ddot{\mathbf{r}}_{O'P} = {}_K\mathbf{F}_a + \underbrace{\left[-m_K\omega \times ({}_K\omega \times {}_K\mathbf{r}_{O'P})\right]}_{\text{centrifugal force}} + \underbrace{\left[-m2_K\omega \times {}_K\dot{\mathbf{r}}_{O'P}\right]}_{\text{Coriolis force}}.$$

These forces have to be added to the momentum equation, if this relation is to be evaluated in a moving system. The Coriolis force will produce a load on the right tube side (negative circumferential direction).

Example 1.6 (Disc on a car). We consider a rocking disc on a car (Fig. 1.12). The rocking function $\varphi(t)$ and the position $x_W(t)$ of the car are given. We want to evaluate the absolute velocity ${}_I\mathbf{v}_A$ and the absolute acceleration ${}_I\mathbf{a}_A$ of the contact point A of the disc. Performing this task, we proceed with the following steps:

- The center point M of the disc can be treated quite easily to give:

$${}_I\mathbf{r}_M = \begin{pmatrix} x_W + L + R\varphi & R & 0 \end{pmatrix}^T.$$

  From this, we get its absolute velocity as

$${}_I\mathbf{v}_M = {}_I\dot{\mathbf{r}}_M = \begin{pmatrix} \dot{x}_W + R\dot{\varphi} & 0 & 0 \end{pmatrix}^T.$$

- With respect to the point A, we apply (1.32). The angular velocity of the disc is ${}_I\omega = \begin{pmatrix} 0 & 0 & -\dot{\varphi} \end{pmatrix}^T$. The relative velocity of A will be zero in a body-fixed frame and as a consequence also in a space-fixed frame after a transformation. The radius vector is ${}_I\mathbf{r}_{MA} = \begin{pmatrix} 0 & -R & 0 \end{pmatrix}^T$, and finally we get

$${}_I\mathbf{v}_A = \begin{pmatrix} \dot{x}_W + R\dot{\varphi} & 0 & 0 \end{pmatrix}^T + \begin{pmatrix} 0 & 0 & -\dot{\varphi} \end{pmatrix}^T \times \begin{pmatrix} 0 & -R & 0 \end{pmatrix}^T = \begin{pmatrix} \dot{x}_W & 0 & 0 \end{pmatrix}^T.$$

  The result corresponds exactly to the *roll condition*.

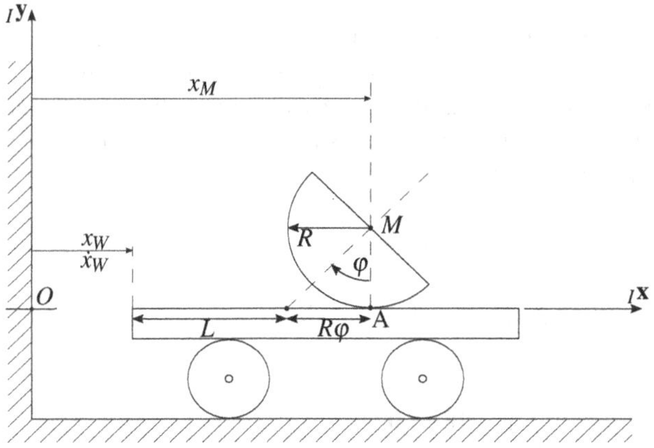

Fig. 1.12. Disc on a car (with the linearization $R\tan(\varphi) \approx R\varphi$).

- The absolute acceleration of the center point M writes

$$_I\mathbf{a}_M = {_I\ddot{\mathbf{r}}_M} = \left(\ddot{x}_W + R\ddot{\varphi}\right).$$

- The relative kinematics of point A follows from the relation (1.36). The relative acceleration of A will also be zero, A being body-fixed. We get

$$\begin{aligned} _I\mathbf{a}_A &= \left(\ddot{x}_W + R\ddot{\varphi}\ 0\ 0\right)^T + \left(0\ 0\ -\ddot{\varphi}\right)^T \times \left(0\ -R\ 0\right)^T \\ &\quad + \left(0\ 0\ -\dot{\varphi}\right)^T \times \left(\left(0\ 0\ -\dot{\varphi}\right)^T \times \left(0\ -R\ 0\right)^T\right) \\ &= \left(\ddot{x}_W + R\ddot{\varphi}\ 0\ 0\right)^T - \left(R\ddot{\varphi}\ 0\ 0\right)^T + \left(0\ 0\ -\dot{\varphi}\right)^T \times \left(-R\dot{\varphi}\ 0\ 0\right)^T \\ &= \left(\ddot{x}_W\ \ R\dot{\varphi}^2\ \ 0\right)^T. \end{aligned}$$

1.5 Momentum and Moment of Momentum

1.5.1 General Axioms

According to Section 1.1, mechanics is interested in those interactions that generate accelerations of bodies or systems of bodies. Therefore, we formulate the very basic axiom that generally external forces on bodies are in equilibrium with the mass or inertia forces [16],

$$\int_K \left(\ddot{\mathbf{r}}\mathrm{d}m - \mathrm{d}\mathbf{F}_a\right) = \mathbf{0}, \tag{1.41}$$

and that all internal forces vanish,

$$\int_K \mathrm{d}\mathbf{F}_i = \mathbf{0}. \tag{1.42}$$

Considering absolute changes with time, this statement is true for an *inertial system* of coordinates (Fig. 1.13), where a radius vector $\mathbf{r}$ points to a mass element $\mathrm{d}m$ loaded by external forces $\mathrm{d}\mathbf{F}_a$. Calculating the torque with respect to point O and keeping in mind the equilibrium statement above, this torque has to be balanced by the moment of momentum in the form:

$$\int_K \mathbf{r} \times \left(\ddot{\mathbf{r}}\mathrm{d}m - \mathrm{d}\mathbf{F}_a\right) = \mathbf{0}. \tag{1.43}$$

This relation obviously can only be true, if at the same time the torque of all internal forces vanishes, which writes

$$\int_K \mathbf{r} \times \mathrm{d}\mathbf{F}_i = \mathbf{0}. \tag{1.44}$$

This requirement turns out to be much more complex than the simple condition that internal forces have to vanish. It anticipates the symmetry of *Cauchy's stress*

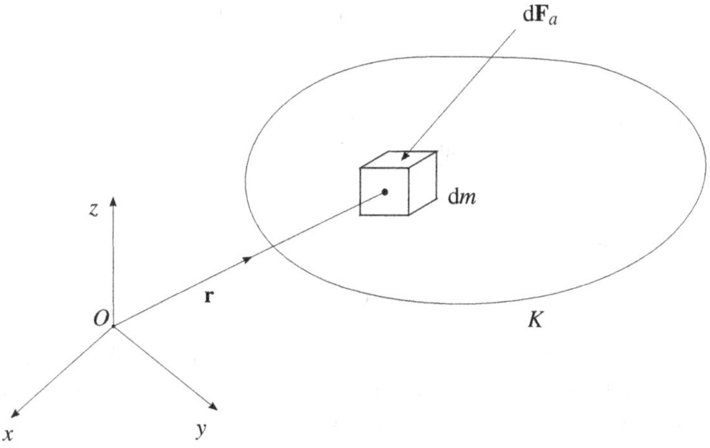

Fig. 1.13. Forces and torques on a mass element.

tensor, which is equivalent to the well-known *axiom of Boltzmann*. Considering the laws of continuum mechanics, the symmetry of Cauchy's stress tensor follows from a moment of momentum balance for a volume element of the continuum [6]. The four postulates as presented above may be considered as axioms resulting from century-old experience. They are sufficient to derive all relations of theoretical and applied mechanics.

1.5.2 Momentum

The equations (1.41), (1.42), and (1.44) also include the three NEWTONian *axioms*.

AXIOM 1: A body at rest remains at rest and a body in motion moves in a straight line with unchanging velocity, unless some external force acts on it.

To illustrate this basic law, which we find already in the statements of GALILEI [67], we use the notation introduced by EULER for the momentum and moment of momentum laws. Referring to this axiom we have no external, thus no active forces, which means $\int_K d\mathbf{F}_a = \mathbf{0}$ and therefore $\int_K \ddot{\mathbf{r}} dm = \mathbf{0}$ resulting in

$$\mathbf{p} = \int_K \dot{\mathbf{r}} dm = \text{const.}, \tag{1.45}$$

which represents the law of conservation of momentum. Considering the mass center of a body, we get

$$\mathbf{p} = \dot{\mathbf{r}}_S m \quad \text{with} \quad \mathbf{r}_S m = \int_K \mathbf{r} dm. \tag{1.46}$$

AXIOM 2:The rate of change of the momentum of a body is proportional to the resultant external force that acts on the body.

For the mass element of Fig. 1.13, we obtain from (1.41)

$$\ddot{\mathbf{r}}\mathrm{dm} - \mathrm{d}\mathbf{F}_a = \mathbf{0}, \tag{1.47}$$

which represents also the momentum budget for a point mass. The time derivative of the momentum is mass times acceleration if we are dealing with a constant mass, as in the above equation. For masses that are not constant the time derivative of the mass must also be considered. In terms of the definitions, we write

$$\frac{\mathrm{d}\mathbf{p}}{\mathrm{d}t} = \mathbf{F}, \quad \text{with} \quad \mathbf{p} = \int_K \dot{\mathbf{r}} \mathrm{d}m, \quad \text{and} \quad \mathbf{F} = \int_K \mathrm{d}\mathbf{F}_a. \tag{1.48}$$

If we consider again the center of mass of the body, we come out with

$$m(\frac{\mathrm{d}\mathbf{v}_S}{\mathrm{d}t}) = \mathbf{F}_S. \tag{1.49}$$

The velocity $\mathbf{v}_S$ is defined with respect to an inertial system. It is an absolute velocity. The force vector $\mathbf{F}_S$ is the vector sum of all forces that act on the body. Generally, this vector sum does not pass through the center of mass resulting in an additional torque, which has to be regarded in the moment of momentum equation.

AXIOM 3: Action and reaction are equal and opposite.

At the time of NEWTON, this finding was new. However, it is very obvious from experience. Wherever any force acts on a body or on the environment, we get as a reaction the same force with opposite sign. My feet transfer my weight to the ground, as a reaction the ground is loaded with my weight force in the opposite direction. There is no mechanical interaction without this basic property.

1.5.3 Moment of Momentum

Leonhard EULER was the first to recognize the moment of momentum equation as an original and independent basic law of dynamics [67]. In agreement with (1.43) and (1.44), we define as *moment of momentum* for a body K

$$\mathbf{L}_O := \int_K \mathbf{r} \times (\dot{\mathbf{r}} \mathrm{d}m) \tag{1.50}$$

and in addition the torque of the external forces

$$\mathbf{M}_O := \int_K \mathbf{r} \times \mathrm{d}\mathbf{F}_a. \tag{1.51}$$

From this, we formulate the following

moment of momentum equation:

$$\frac{\mathrm{d}\mathbf{L}_O}{\mathrm{d}t} = \mathbf{M}_O. \tag{1.52}$$

We have to keep in mind that the moment of momentum as well as the torque depend always on some reference point. As we know from the discussion above, the definition of the moment of momentum equation (1.52) presupposes Boltzmann's axiom or, equivalently, the symmetry of Cauchy's stress tensor [6]. Deriving the moment of momentum with respect to time requires EULER's differentiation rule (Section 1.4.3) [41].

1.6 Energy

A mass element $\mathrm{d}m$ moving by $\mathrm{d}\mathbf{r}$ from a point 1 to a point 2 in a *force field* (Fig. 1.14) produces *physical work*. This work might be expressed either by the external forces or by the equivalent acceleration (1.41). It is

$$\mathrm{d}W = \int_{\mathbf{r}_1}^{\mathbf{r}_2} \mathrm{d}\mathbf{F}_a^T \mathrm{d}\mathbf{r} = \mathrm{d}m \int_{\mathbf{r}_1}^{\mathbf{r}_2} \ddot{\mathbf{r}}^T \mathrm{d}\mathbf{r}. \tag{1.53}$$

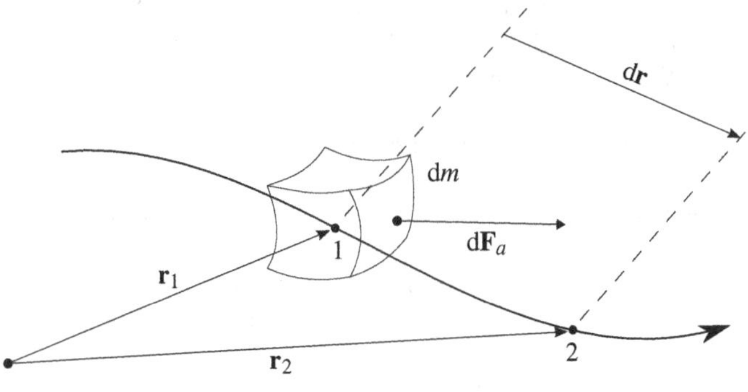

Fig. 1.14. Mass element in a field of forces.

The acceleration term can be manipulated in the following form

$$\mathrm{d}m \int_{\mathbf{r}_1}^{\mathbf{r}_2} \ddot{\mathbf{r}}^T \mathrm{d}\mathbf{r} = \mathrm{d}m \int_{\mathbf{r}_1}^{\mathbf{r}_2} \frac{\mathrm{d}\dot{\mathbf{r}}^T}{\mathrm{d}t} \mathrm{d}\mathbf{r} = \mathrm{d}m \int_{\dot{\mathbf{r}}_1}^{\dot{\mathbf{r}}_2} \dot{\mathbf{r}}^T \mathrm{d}\dot{\mathbf{r}} \tag{1.54}$$

resulting after an integration in a dependency on the *kinetic energy*

$$\mathrm{d}m \int_{\mathbf{r}_1}^{\mathbf{r}_2} \ddot{\mathbf{r}}^T \mathrm{d}\mathbf{r} = \frac{1}{2} \mathrm{d}m \left[\dot{\mathbf{r}}_2^2 - \dot{\mathbf{r}}_1^2\right] =: \mathrm{d}T_2 - \mathrm{d}T_1. \tag{1.55}$$

From this, we conclude, that the work done by the external forces $\mathrm{d}\mathbf{F}_a$ is equal to the difference between the corresponding kinetic energies.

Energy equation

$$\mathrm{d}W = \mathrm{d}T_2 - \mathrm{d}T_1. \tag{1.56}$$

We now focus our considerations on fields of forces, where along a closed trajectory Γ no work is done. Examples are gravitational fields, the electric fields of COULOMB, all central force fields, and also the forces of springs. This means that

$$\oint_\Gamma \mathrm{d}\mathbf{F}_a^T \mathrm{d}\mathbf{r} = 0. \tag{1.57}$$

Systems with such features are called *conservative systems*. They are free from energy losses and also from energy sources. Applying STOKES' theorem [77] to (1.57), we get

$$\oint_\Gamma \mathrm{d}\mathbf{F}_a^T \mathrm{d}\mathbf{r} = \int_A \nabla \times \mathrm{d}\mathbf{F}_a \mathrm{d}A. \tag{1.58}$$

The area $\mathrm{d}A$ is surrounded by the closed trajectory Γ, where $\nabla \times \mathrm{d}\mathbf{F}_a$ is the *rotation of the force field* through this area. With $\mathrm{d}A$ being arbitrary, we arrive at the *irrotationality of the force field* in the form

$$\nabla \times \mathrm{d}\mathbf{F}_a = \mathbf{0}, \tag{1.59}$$

which is equivalent to (1.57). The rotation of a flow field corresponds to twice its angular velocity.

Both relations (1.57) and (1.59) are necessary conditions for the existence of a *potential*; for practical applications they are also sufficient conditions [58]:

$$\mathrm{d}\mathbf{F}_a^T =: -\nabla \mathrm{d}V. \tag{1.60}$$

The magnitude $\mathrm{d}V$ is called the *potential energy* of the mass element under consideration, sometimes also its *potential function*. From the above equations, we obtain

$$\int_{\mathbf{r}_1}^{\mathbf{r}_2} \mathrm{d}\mathbf{F}_a^T \mathrm{d}\mathbf{r} = -\int_{\mathbf{r}_1}^{\mathbf{r}_2} \left(\frac{\partial \mathrm{d}V}{\partial \mathbf{r}}\right)^T \mathrm{d}\mathbf{r} = -\int_{\mathrm{d}V_1}^{\mathrm{d}V_2} \mathrm{d}(\mathrm{d}V) = -(\mathrm{d}V_2 - \mathrm{d}V_1). \tag{1.61}$$

Therefore, the work $\mathrm{d}W$ is identical with the negative difference of the potential $\mathrm{d}V$:

$$\mathrm{d}W = -(\mathrm{d}V_2 - \mathrm{d}V_1) = \mathrm{d}T_2 - \mathrm{d}T_1. \tag{1.62}$$

Considering these relations and rearranging a bit we see, that for conservative systems the total energy at point 1 must be the same as for point 2:

$$\mathrm{d}E_1 := \mathrm{d}T_1 + \mathrm{d}V_1 = \mathrm{d}T_2 + \mathrm{d}V_2 =: \mathrm{d}E_2. \tag{1.63}$$

For a total mass m and not only a mass element $\mathrm{d}m$, we get the

energy equation for conservative systems:

$$T + V = \text{const.}. \tag{1.64}$$

For conservative systems, the energy budget does not change. However, for example having friction in the system, the total energy would decrease and the line integral (1.57) would not vanish. Force fields depending on time and/or velocity are not conservative. The discussion of energy results more or less from the discussion of momentum and moment of momentum, including Boltzmann's axiom. From this, the energy statements are not an additional axiom. However they are, fortunately, a first *integral of motion*, at least for conservative systems, and thus an invariant magnitude of motion. This can be very useful with respect to further considerations of dynamics.

The relations for momentum and moment of momentum are sufficient for analyzing the motion of any mechanical system. However, in many cases their application is costly and sometimes difficult. For most problems of practical relevancy, we need to deal with constraints, which makes another approach necessary. We come back to this immediately. Anyway, in connection with these principles, the merit of Isaac NEWTON (1642-1727) consists in the formulation of the three basic axioms of momentum, and the merit of EULER (1707-1783) in the introduction of the cut principle and a concise formulation of the moment of momentum axiom as an original and independent law.

1.7 Principles of d'Alembert and Jourdain

1.7.1 Significance of Constraints

In this section we introduce the following concept of forces [50]. The pair "impressed" and "constraint (reaction)" forces describes in the first case given or applied forces, very often interpretable as physical laws (gravitation, magnetic fields, springs and dampers etc.), and in the second case forces produced by constraints. The pair "active" and "passive" forces describes forces that contribute or do not contribute to motion. Reaction forces are always passive forces due to their constraining character. Impressed forces may be active or passive [51]. As already discussed in Section 1.3, the definition of internal and external forces depends on the structure of the free body diagram, that is how we have designed our cuts.

The existence of constraints implies two difficulties. The first one concerns the independence of coordinates, which are constrained. Therefore, the original coordinate set, for example in some three-dimensional workspace, does not represent the possible number of degrees of freedom. Some of the equations of motion depend on each other. The second difficulty is connected with the forces due to constraints. These constraint forces are not given a priori, they must be evaluated by the solution process. Moreover, as the constraint forces do not contribute to the motion of the system, they are reaction forces holding the system together, where we should bear in mind that passive forces and motion means passive forces and relative motion. From the technical viewpoint, we need them as forces in bearings, guides, joints, and the like; they determine system design. The inverse is also true: every system design defines constraints.

The derivation of the equations of motion in respect of constraints should take place with the requirements that

- the constraints can be taken into account by some automatic procedure,
- it is as simple and transparent as possible,
- the resulting differential equations can easily be solved.

These requirements define something like the squaring of a circle, but they also define a direction to go in. The most effective tools for considering dynamic systems with constraints are the principles of mechanics, like those of d'ALEMBERT, JOURDAIN, GAUSS, and HAMILTON [10, 16, 28, 59, 68], to give a few and most important examples. We start with the principles of d'ALEMBERT and JOURDAIN.

1.7.2 Principle of d'Alembert

Summing up (1.41) and (1.42) by considering a body K, partitioning additionally the forces into *impressed and constraint forces* (instead of external and internal forces), we get

$$\int_K \ddot{\mathbf{r}} \mathrm{d}m = \int_K \mathrm{d}\mathbf{F} = \int_K (\mathrm{d}\mathbf{F}^e + \mathrm{d}\mathbf{F}^z). \tag{1.65}$$

Applying a virtual displacement $\delta\mathbf{r}$ to the mass element $\mathrm{d}m$ (Fig. 1.15), we come out with

$$(\ddot{\mathbf{r}} \mathrm{d}m - \mathrm{d}\mathbf{F}^e)^T \, \delta\mathbf{r} = (\mathrm{d}\mathbf{F}^z)^T \, \delta\mathbf{r} = \delta W^z. \tag{1.66}$$

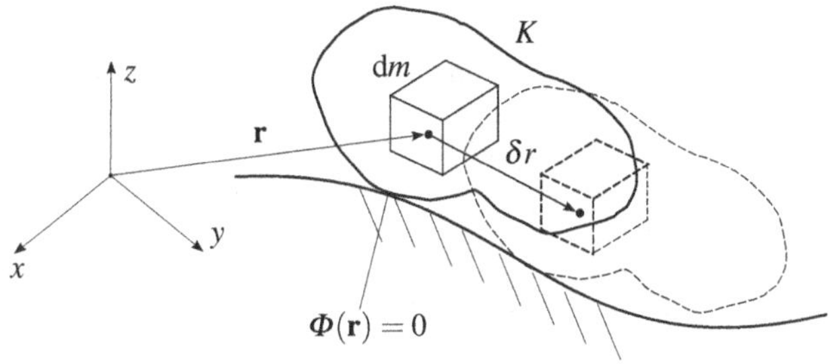

Fig. 1.15. Virtual displacement of a mass element.

This is the *virtual work* as a result of the virtual displacement $\delta\mathbf{r}$. It is generated either by the impressed forces $d\mathbf{F}^e$ and the mass forces $\ddot{\mathbf{r}}dm$ or by the constraint forces $d\mathbf{F}^z$. Without loss of generality, we assume holonomic-scleronomic constraints in the following:

$$\boldsymbol{\Phi}(\mathbf{r}) = \mathbf{0} \quad \text{with} \quad \mathbf{r} \in \mathrm{IR}^3, \quad \boldsymbol{\Phi} \in \mathrm{IR}^m, \quad m < 3. \tag{1.67}$$

The constraint equations do not allow a free choice of the virtual displacements, which always have to be compatible with these constraints. We write (Section 1.3.4)

$$\mathbf{0} \overset{!}{=} \boldsymbol{\Phi}(\mathbf{r} + \delta\mathbf{r}) \doteq \underbrace{\boldsymbol{\Phi}(\mathbf{r})}_{=\mathbf{0}} + \frac{\partial \boldsymbol{\Phi}}{\partial \mathbf{r}} \delta\mathbf{r} =: \delta\boldsymbol{\Phi}. \tag{1.68}$$

This relation means geometrically, that m constraints (1.67) in the space IR^3 span in total m constraint surfaces $\{\Phi_\nu(\mathbf{r}) = 0\}_{\nu=1}^m$, the surface normals $\mathbf{n}_\nu$ of which are proportional to $\left(\frac{\partial \Phi_\nu}{\partial \mathbf{r}}\right)^T$, see Fig. 1.16:

$$\left(\frac{\partial \Phi_\nu}{\partial \mathbf{r}}\right)^T \perp \delta\mathbf{r}. \tag{1.69}$$

Any virtual displacement can only take place on the surfaces of constraints, not perpendicular to them. Conversely, the constraint forces as generated by the constraints $\boldsymbol{\Phi}(\mathbf{r}) = \mathbf{0}$ can only be perpendicular to these surfaces, which means in the normal surface direction (Section 1.3.3), because the situation where any body leaves the constraint surfaces must be avoided. It is a very hard condition, that free motion must and only can take place within these surfaces. The term $(d\mathbf{F}^z)^T \delta\mathbf{r} = 0$ represents the substantial meaning of d'ALEMBERT's principle, which writes using LAGRANGE's version [28, 68]:

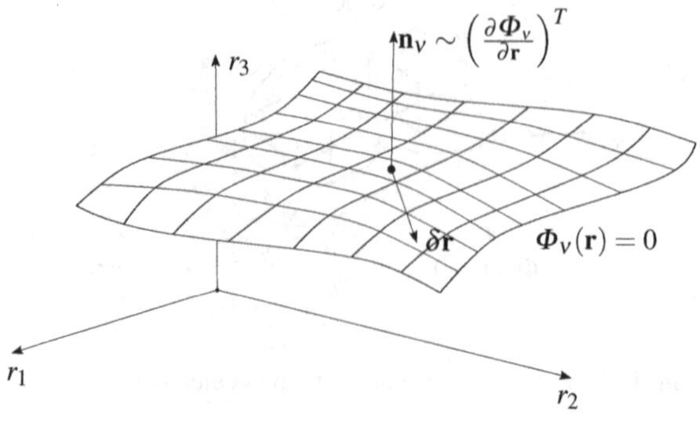

Fig. 1.16. Constraint surfaces.

constraint forces do no work,

$$\int_K (\mathrm{d}\mathbf{F}^z)^T\, \delta\mathbf{r} = 0. \tag{1.70}$$

With respect to generating motion they are "lost forces", which compel the dynamic system into the surfaces of constraints. The motion itself will only be produced by the impressed forces.

Principle of d'ALEMBERT:

$$\int_K (\ddot{\mathbf{r}}\mathrm{d}m - \mathrm{d}\mathbf{F}^e)^T\, \delta\mathbf{r} = 0. \tag{1.71}$$

It should be noted that constraint forces may be shifted in the tangential plane of the constraint surfaces without violating conditions (1.69) and (1.70), because these conditions only say that the constraint forces $\mathrm{d}\mathbf{F}^z$ must be perpendicular to $\delta\mathbf{r}$. This property is important for problems including friction, where for COULOMB's friction law the absolute value of sliding friction is proportional to the absolute value of the constraint force in normal direction of a contact.

With respect to the static case ($\ddot{\mathbf{r}} = 0$), D'ALEMBERT's principle (1.71) reduces to the *principle of virtual work* [41]:

$$\delta W = \int_K (\mathrm{d}\mathbf{F})^T\, \delta\mathbf{r} = 0. \tag{1.72}$$

According to D'ALEMBERT's principle, we may split up the original relation

$$\int_K (\ddot{\mathbf{r}}\mathrm{d}m - \mathrm{d}\mathbf{F}^e) = \int_K \mathrm{d}\mathbf{F}^z \tag{1.73}$$

into two terms, one being true for the directions perpendicular to the constraint surfaces $\boldsymbol{\Phi}(\mathbf{r}) = \mathbf{0}$ ($\perp$), and the other being true for all directions tangential to the constraint surfaces ($\|$):

$$\int_K (\ddot{\mathbf{r}}\mathrm{d}m - \mathrm{d}\mathbf{F}^e)_\perp + \int_K (\ddot{\mathbf{r}}\mathrm{d}m - \mathrm{d}\mathbf{F}^e)_\| = \int_K (\mathrm{d}\mathbf{F}^z)_\perp + \int_K (\mathrm{d}\mathbf{F}^z)_\| . \tag{1.74}$$

As the motion definitely takes place only on the surfaces $\boldsymbol{\Phi}(\mathbf{r}) = \mathbf{0}$, we require that the normal and tangential parts become zero independently:

$$\int_K (\ddot{\mathbf{r}}\mathrm{d}m - \mathrm{d}\mathbf{F}^e)_\| = \int_K (\mathrm{d}\mathbf{F}^z)_\| = \mathbf{0}, \tag{1.75}$$

$$\int_K (\ddot{\mathbf{r}}\mathrm{d}m - \mathrm{d}\mathbf{F}^e - \mathrm{d}\mathbf{F}^z)_\perp = \mathbf{0}, \tag{1.76}$$

which means:

- The impressed tangential forces alone contribute to any acceleration of masses.
- The normal components of the impressed forces are in static equilibrium with the constraint forces and the mass forces $(\ddot{\mathbf{r}}\mathrm{d}m)_\perp$.

We close this section with some historical remarks concerning the evolution of the principle of D'ALEMBERT [67]:

1. Daniel BERNOULLI (1700 - 1782)
   Forces not contributing to accelerations are "lost forces":

$$\ddot{\mathbf{r}}\mathrm{d}m - \mathrm{d}\mathbf{F}^e = \mathrm{d}\mathbf{F}^z.$$

2. Jean-Baptiste le Rond d'ALEMBERT (1717 - 1783)
   The sum of all "lost forces" must be in an equilibrium state.
3. Joseph-Louis LAGRANGE (1736 - 1813)
   "Lost forces" do no work. From this we have the so-called LAGRANGE's version:

$$(\ddot{\mathbf{r}}\mathrm{d}m - \mathrm{d}\mathbf{F}^e)^T \,\delta\mathbf{r} = (\mathrm{d}\mathbf{F}^z)^T \,\delta\mathbf{r} = 0.$$

1.7.3 Principle of Jourdain

The principle of JOURDAIN follows qualitatively from the same arguments that we used for the principle of D'ALEMBERT in the form of LAGRANGE. Constraint equations define a certain type of motion, they give an instruction as to how motion should develop. A revolute joint for example prescribes at its location an angular motion with one degree of freedom only. The constraint forces resulting from such a constraint are always perpendicular to the surface as defined by the corresponding kinematic magnitudes of the constraint. They cannot be displaced in their positive direction, and constraint torques cannot be rotated. Thus, the example of the revolute joint requires that the corresponding constraint force is perpendicular to the joint axis.

From these properties we conclude that such constraint forces not only do no work according to D'ALEMBERT (1.70), but that they also cannot produce power. This is the substantial statement of JOURDAIN's principle. The scalar product of the vector $d\mathbf{F}^z$ of constraint forces and the vector $\delta\dot{\mathbf{r}}$ of nearly any *virtual velocity* must vanish.

Philip JOURDAIN (1879 - 1919):
Constraint forces generate no power.

$$\int_K (d\mathbf{F}^z)^T\,\delta\dot{\mathbf{r}} = 0. \tag{1.77}$$

The principle of JOURDAIN is obviously very advantageous with respect to nonholonomic constraints, because we can use them directly. Holonomic constraints (1.7) must be differentiated formally to give

$$\boldsymbol{\Phi}(\mathbf{r},\dot{\mathbf{r}},t) = \mathbf{W}(\mathbf{r},t)^T\,\dot{\mathbf{r}} + \hat{\mathbf{w}}(\mathbf{r},t) = \mathbf{0} \quad \text{with} \quad \dot{\mathbf{r}} \in \mathrm{IR}^3, \quad \boldsymbol{\Phi} \in \mathrm{IR}^m, \quad m < 3. \tag{1.78}$$

D'ALEMBERT's principle is limited to holonomic constraints and small virtual displacements. JOURDAIN's principle is much more general including all possible constraints, if we allow formal differentiation. We consider changes concerning the virtual velocity, $\delta\dot{\mathbf{r}}$, $\delta\mathbf{r} = \mathbf{0}$, $\delta t = 0$ [50], which results in

$$\mathbf{0} \stackrel{!}{=} \boldsymbol{\Phi}(\mathbf{r},\dot{\mathbf{r}}+\delta\dot{\mathbf{r}},t) = \underbrace{\boldsymbol{\Phi}(\mathbf{r},\dot{\mathbf{r}},t)}_{=\mathbf{0}} + \frac{\partial\boldsymbol{\Phi}}{\partial\dot{\mathbf{r}}}\delta\dot{\mathbf{r}} =: \delta\boldsymbol{\Phi}. \tag{1.79}$$

This again confirms the same properties as before; the virtual velocities and the constraint normal must be perpendicular to each other:

$$\mathbf{W} \perp \delta\dot{\mathbf{r}} \qquad \text{with} \qquad \mathbf{W} = \frac{\partial\boldsymbol{\Phi}}{\partial\dot{\mathbf{r}}}. \tag{1.80}$$

We derived (1.78) by formal differentiation of some holonomic constraint resulting in a linear relationship with respect to velocity. This is of much more general significance, because additionally all physically meaningful nonholonomic constraints always possess this linear form. Up to now no real exception has been discerned. However, this means that the variation of the velocity $\partial\dot{\mathbf{r}}$ in JOURDAIN's principle (1.77) need not be small, necessarily. In summation, we can write

$$\int_K (\ddot{\mathbf{r}}\mathrm{d}m - \mathrm{d}\mathbf{F}^e)^T \,\delta\dot{\mathbf{r}} = 0. \tag{1.81}$$

1.8 Newton-Euler Equations for Constrained Systems

The really brilliant principles of D'ALEMBERT and JOURDAIN offer an elegant and transparent possibility to combine the equations of motion of a single body or of a system of bodies with all constraints concerning such a mass system. We consider the most convenient methods in the following.

1.8.1 Single Rigid Body

The relationship (1.81) includes the absolute velocity $\dot{\mathbf{r}}$ and the absolute acceleration $\ddot{\mathbf{r}}$ of an arbitrary point of a rigid body K. For expressing the absolute magnitudes by relative ones we define this arbitrary point as point P. We obtain from (1.32) and (1.36) together with the rigid body assumptions ${}_K\dot{\mathbf{r}}_{O'P} = \mathbf{0}$ and ${}_K\ddot{\mathbf{r}}_{O'P} = \mathbf{0}$

$$\dot{\mathbf{r}} = \mathbf{v}_{O'} + \tilde{\omega}\mathbf{r}_{O'P}, \tag{1.82}$$

$$\ddot{\mathbf{r}} = \mathbf{a}_{O'} + \left(\dot{\tilde{\omega}} + \tilde{\omega}\tilde{\omega}\right)\mathbf{r}_{O'P}. \tag{1.83}$$

Provided we find a set of minimal coordinates $\mathbf{q}$, $\dot{\mathbf{q}}$, we express $\mathbf{r}, \dot{\mathbf{r}}$ by the functions $\mathbf{r}(\mathbf{q},t)$ and $\dot{\mathbf{r}}(\mathbf{q},\dot{\mathbf{q}},t)$. These minimal coordinates satisfy all constraints (1.78) automatically. The time derivation $\dot{\mathbf{r}}$ can be written in the form

$$\dot{\mathbf{r}} = \left(\frac{\partial \mathbf{r}}{\partial \mathbf{q}}\right)\dot{\mathbf{q}} + \frac{\partial \mathbf{r}}{\partial t}. \tag{1.84}$$

From this, we come to the very important relation

$$\left(\frac{\partial \dot{\mathbf{r}}}{\partial \dot{\mathbf{q}}}\right) = \left(\frac{\partial \mathbf{r}}{\partial \mathbf{q}}\right). \tag{1.85}$$

According to the agreement above, we consider only $\delta\dot{\mathbf{q}}$-variations and get

$$\delta\dot{\mathbf{r}} = \left(\frac{\partial \dot{\mathbf{r}}}{\partial \dot{\mathbf{q}}}\right)\delta\dot{\mathbf{q}} = \left(\frac{\partial \mathbf{r}}{\partial \mathbf{q}}\right)\delta\dot{\mathbf{q}}. \tag{1.86}$$

Regarding (1.82), the relation $\tilde{\omega}\mathbf{r}_{O'P} = \tilde{\mathbf{r}}^T_{O'P}\omega$ gives

$$\delta\dot{\mathbf{r}} = \left[\left(\frac{\partial \mathbf{v}_{O'}}{\partial \dot{\mathbf{q}}}\right) + \tilde{\mathbf{r}}^T_{O'P}\left(\frac{\partial \omega}{\partial \dot{\mathbf{q}}}\right)\right]\delta\dot{\mathbf{q}}. \tag{1.87}$$

Composing this relation together with (1.83) and (1.81) we derive the following integral

$$\int_K \delta\dot{\mathbf{q}}^T \left[\left(\frac{\partial \mathbf{v}_{O'}}{\partial \dot{\mathbf{q}}}\right) + \tilde{\mathbf{r}}_{O'P}^T \left(\frac{\partial \omega}{\partial \dot{\mathbf{q}}}\right)\right]^T \left\{\left[\mathbf{a}_{O'} + (\dot{\tilde{\omega}} + \tilde{\omega}\tilde{\omega})\,\mathbf{r}_{O'P}\right] dm - \mathrm{d}\mathbf{F}^e\right\} = 0. \tag{1.88}$$

Now, we choose the virtual velocities $\delta\dot{\mathbf{q}}$ in an arbitrary way, because the minimal coordinates satisfy the constraints automatically. Applying the *fundamental lemma of variational calculus* to the equation above, which requires the integrand to be zero, and arranging the terms according to $\left(\frac{\partial \mathbf{v}_{O'}}{\partial \dot{\mathbf{q}}}\right)^T$ and $\left(\frac{\partial \omega}{\partial \dot{\mathbf{q}}}\right)^T$, we finally get

$$\begin{aligned}
\mathbf{0} = &\left(\frac{\partial \mathbf{v}_{O'}}{\partial \dot{\mathbf{q}}}\right)^T \left[\int_K (\mathbf{a}_{O'} + (\dot{\tilde{\omega}} + \tilde{\omega}\tilde{\omega})\,\mathbf{r}_{O'P})\,\mathrm{d}m - \int_K \mathrm{d}\mathbf{F}^e\right] \\
&+ \left(\frac{\partial \omega}{\partial \dot{\mathbf{q}}}\right)^T \left[\int_K \tilde{\mathbf{r}}_{O'P}\mathbf{a}_{O'}\mathrm{d}m + \int_K \tilde{\mathbf{r}}_{O'P}\dot{\tilde{\omega}}\mathbf{r}_{O'P}\mathrm{d}m + \int_K \tilde{\mathbf{r}}_{O'P}\tilde{\omega}\tilde{\omega}\mathbf{r}_{O'P}\mathrm{d}m - \int_K \tilde{\mathbf{r}}_{O'P}\mathrm{d}\mathbf{F}^e\right].
\end{aligned} \tag{1.89}$$

These are the momentum and the moment of momentum equations projected into the free directions of motion, which means, into the tangential directions of the constraint surfaces. We return to the concept of impressed forces (Section 1.5)

$$\mathbf{F}^e = \int_K \mathrm{d}\mathbf{F}^e, \tag{1.90}$$

$$\mathbf{M}_{O'}^e = \int_K \tilde{\mathbf{r}}_{O'P}\,\mathrm{d}\mathbf{F}^e \tag{1.91}$$

and apply the definitions of mass center (1.46)

$$\mathbf{r}_{O'S} = \frac{1}{m}\int_K \mathbf{r}_{O'P}\,\mathrm{d}m \tag{1.92}$$

and inertia tensor

$$\Theta_{O'} = -\int_K \tilde{\mathbf{r}}_{O'P}\tilde{\mathbf{r}}_{O'P}\,\mathrm{d}m, \tag{1.93}$$

which can be referred to the mass center by the *rule of* HUYGENS-STEINER

$$\begin{aligned}
\Theta_S &= -\int_K \tilde{\mathbf{r}}_{SP}\tilde{\mathbf{r}}_{SP}\,\mathrm{d}m = -\int_K (\tilde{\mathbf{r}}_{SO'} + \tilde{\mathbf{r}}_{O'P})(\tilde{\mathbf{r}}_{SO'} + \tilde{\mathbf{r}}_{O'P})\,\mathrm{d}m \\
&= \Theta_{O'} - m\tilde{\mathbf{r}}_{O'S}^2 - \int_K \tilde{\mathbf{r}}_{SO'}\tilde{\mathbf{r}}_{O'P}\,\mathrm{d}m - \int_K \tilde{\mathbf{r}}_{O'P}\tilde{\mathbf{r}}_{SO'}\,\mathrm{d}m \\
&= \Theta_{O'} + m\tilde{\mathbf{r}}_{O'S}^2.
\end{aligned} \tag{1.94}$$

We finally obtain

$$\mathbf{0} = \mathbf{J}_{O'}^{T}\left[m\mathbf{a}_{O'} + m\left(\dot{\tilde{\omega}} + \tilde{\omega}\tilde{\omega}\right)\mathbf{r}_{O'S} - \mathbf{F}^{e}\right] + \mathbf{J}_{R}^{T}\left[\underbrace{m\tilde{\mathbf{r}}_{O'S}\mathbf{a}_{O'}}_{\text{orbital influence}} + \underbrace{\Theta_{O'}\dot{\omega}}_{\text{relative influence}} + \underbrace{\tilde{\omega}\Theta_{O'}\omega}_{\text{centrifugal influence}} - \mathbf{M}_{O'}^{e}\right] \quad (1.95)$$

with the JACOBIANs

$$\mathbf{J}_{O'} = \left(\frac{\partial \mathbf{v}_{O'}}{\partial \dot{\mathbf{q}}}\right), \quad (1.96)$$

$$\mathbf{J}_{R} = \left(\frac{\partial \omega}{\partial \dot{\mathbf{q}}}\right). \quad (1.97)$$

For the manipulation of the centrifugal influences we used the JACOBI identity

$$\mathbf{i} \times (\mathbf{j} \times \mathbf{k}) = -\mathbf{j} \times (\mathbf{k} \times \mathbf{i}) - \mathbf{k} \times (\mathbf{i} \times \mathbf{j}) \quad (1.98)$$

with $\mathbf{i} = \mathbf{r}_{O'P}$, $\mathbf{j} = \omega$, $\mathbf{k} = \omega \times \mathbf{r}_{O'P}$.

1.8.2 System with Multiple Rigid Bodies

The equation (1.95) can easily be extended to a *multibody system (MBS)* with many rigid bodies by summing up all n bodies of the multibody system. All magnitudes are indexed by i, and (1.95) becomes

$$\sum_{i=1}^{n}\mathbf{J}_{O'_i}^{T}\left[m_i\mathbf{a}_{O'_i} + m_i\left(\dot{\tilde{\omega}}_i + \tilde{\omega}_i\tilde{\omega}_i\right)\mathbf{r}_{O'_iS_i} - \mathbf{F}_i^{e}\right] + \sum_{i=1}^{n}\mathbf{J}_{R_i}^{T}\left[m_i\tilde{\mathbf{r}}_{O'_iS_i}\mathbf{a}_{O'_i} + \Theta_{O'_i,i}\dot{\omega}_i + \tilde{\omega}_i\Theta_{O'_i,i}\omega_i - \mathbf{M}_{O'_i,i}^{e}\right] = \mathbf{0}. \quad (1.99)$$

1.8.3 Remarks

1. By choosing the mass center of bodies as reference points, which in many cases will be possible, the momentum and moment of momentum equations simplify considerably:

$$\mathbf{0} = \mathbf{J}_{S}^{T}\left[m\mathbf{a}_{S} - \mathbf{F}^{e}\right] + \mathbf{J}_{R}^{T}\left[\Theta_{S}\dot{\omega} + \tilde{\omega}\Theta_{S}\omega - \mathbf{M}_{S}^{e}\right]. \quad (1.100)$$

2. What is the principal meaning of the relations (1.95)? The row vectors of the JACOBIANs $\mathbf{J}_{O'}$, $\mathbf{J}_{R}$ represent the gradients with respect to $\dot{\mathbf{q}}$ of the constraint surfaces given by $\mathbf{v}_{O'}(\mathbf{q}, \dot{\mathbf{q}}, t)$ and $\omega(\mathbf{q}, \dot{\mathbf{q}}, t)$. The multiplication of the JACOBIAN

of translation $\mathbf{J}_{O'}$ with the momentum equation and that of the JACOBIAN of rotation $\mathbf{J}_R$ with the moment of momentum equation describes the projection to the constraint surfaces as discussed in Section 1.7.2 (Fig. 1.16). Therefore, the relation (1.95) cuts out all those forces and torques, which are "lost forces" with respect to the motion. Only those forces and torques are taken into consideration that contribute to acceleration of the body into the free directions allowed by the constraints.

3. For every row of $\mathbf{J}_{O'}^T, \mathbf{J}_R^T$, we have a scalar product with the corresponding momentum or moment of momentum equation in (1.95), and of course also in (1.99). Each of these scalar products may be evaluated in different coordinate systems without changing the result, provided that the two vector components of such a scalar product are defined in the same coordinate system. This simplifies the evaluation considerably, because we could evaluate the momentum equation in a space-fixed frame, and the moment of momentum equation in a body-fixed frame with a constant inertia tensor (1.93), for example.

4. For the free motion of a single rigid body without constraints, we may choose $\mathbf{q} = \mathbf{r}$, indicating that the coupling of translation and rotation vanishes. The corresponding equations are (Section 1.5).

$$m\mathbf{a}_{O'} + m\left(\dot{\tilde{\omega}} + \tilde{\omega}\tilde{\omega}\right)\mathbf{r}_{O'S} = \mathbf{F}^e, \tag{1.101}$$

$$m\tilde{\mathbf{r}}_{O'S}\mathbf{a}_{O'} + \Theta_{O'}\dot{\omega} + \tilde{\omega}\Theta_{O'}\omega = \mathbf{M}^e_{O'} \tag{1.102}$$

5. The second equation above represents the moment of momentum equation in the same form as that of the second part of (1.95), namely given with respect to a moving body-fixed point. Conversely, (1.52) refers to an inertial system. To understand the relationship between both representations, we go back to (1.50) and take into account additionally the relations (1.22) and (1.82). We obtain

$$\mathbf{L}_O := \int_K \mathbf{r} \times (\dot{\mathbf{r}}dm) = \mathbf{L}_{O'} + m\left[\mathbf{r}_{O'} \times (\mathbf{v}_{O'} + \omega \times \mathbf{r}_{O'S}) + \mathbf{r}_{O'S} \times \mathbf{v}_{O'}\right]. \tag{1.103}$$

Considering EULER's theorem, the momentum equation, and some standard rules for the vector product, we modify the second term to give the absolute inertial change

$$\begin{aligned}
&(m\mathbf{v}_{O'}) \times (\mathbf{v}_{O'} + \omega \times \mathbf{r}_{O'S}) + \mathbf{r}_{O'} \times \left(m\mathbf{a}_{O'} + m(\dot{\tilde{\omega}} + \tilde{\omega}\tilde{\omega})\mathbf{r}_{O'S}\right) \\
&\quad + (\omega \times \mathbf{r}_{O'S}) \times (m\mathbf{v}_{O'}) + (m\mathbf{r}_{O'S}) \times \mathbf{a}_{O'} \\
&= \mathbf{r}_{O'} \times \mathbf{F}^e + (m\mathbf{r}_{O'S}) \times \mathbf{a}_{O'}.
\end{aligned} \tag{1.104}$$

Taking the torque equation into account

$$\mathbf{M}_O = \mathbf{M}_{O'} + \mathbf{r}_{O'} \times \mathbf{F}^e, \tag{1.105}$$

we realize that the moment of momentum with respect to an inertial frame (1.52) follows from the moment of momentum with respect to a body-fixed frame (1.95) with moving reference O', or vice versa.

6. Starting with the absolute changes $(m\,{}_K\mathbf{a}_{O'} + m\left({}_K\dot{\tilde{\omega}} + {}_K\tilde{\omega}\,{}_K\tilde{\omega}\right){}_K\mathbf{r}_{O'S})$ of the momentum and the absolute changes $({}_K\Theta_{O'}\,{}_K\dot{\omega} + {}_K\tilde{\omega}\,{}_K\Theta_{O'}\,{}_K\omega)$ of the moment of momentum represented in a body-fixed frame with reference O', we may reproduce momentum and moment of momentum by EULER's theorem to give

$$ {}_K\mathbf{p} = m\,{}_K\mathbf{v}_{O'} + m\,{}_K\tilde{\omega}\,{}_K\mathbf{r}_{O'S}\,, \tag{1.106} $$
$$ {}_K\mathbf{L}_{O'} = {}_K\Theta_{O'}\,{}_K\omega\,. \tag{1.107} $$

For an arbitrary reference frame, we evaluate momentum and moment of momentum dependent on the relevant angular velocities by

$$ {}_R\omega_R \times {}_R\mathbf{p} + {}_R\dot{\mathbf{p}}\,, \tag{1.108} $$
$$ {}_R\omega_R \times {}_R\mathbf{L}_{O'} + {}_R\dot{\mathbf{L}}_{O'}\,. \tag{1.109} $$

7. Assuming that $O' = S$ is the mass center with $(\mathbf{r}_{O'S} = \mathbf{0})$ or that alternatively O' is an inertial body-fixed point $(\mathbf{a}_{O'} = \mathbf{0})$, the moment of momentum equation becomes

$$ \Theta_{O'}\,\dot{\omega} + \omega \times \Theta_{O'}\,\omega = \mathbf{M}^e_{O'}\,, \tag{1.110} $$

which corresponds to the vector form of the *dynamic* EULER *equation*, if we assume a principal coordinate system [41, 23]. It is an important tool of gyrodynamics.

8. The JACOBIAN of rotation has been defined by (1.97)

$$ \mathbf{J}_R = \left(\frac{\partial\omega}{\partial\dot{\mathbf{q}}}\right). \tag{1.111} $$

The JACOBIAN of translation writes (1.85):

$$ \mathbf{J}_T = \left(\frac{\partial\dot{\mathbf{r}}}{\partial\dot{\mathbf{q}}}\right) = \left(\frac{\partial\mathbf{r}}{\partial\mathbf{q}}\right). \tag{1.112} $$

According to Section 1.4, we know

$$ \omega = \mathbf{Y}^{-1}(\bar{\varphi})\dot{\bar{\varphi}}. \tag{1.113} $$

With $\bar{\varphi} = \bar{\varphi}(\mathbf{q},t)$, it is

$$ \dot{\bar{\varphi}} = \left(\frac{\partial\bar{\varphi}}{\partial\mathbf{q}}\right)\dot{\mathbf{q}} + \frac{\partial\bar{\varphi}}{\partial t} \tag{1.114} $$

and analogously also here

$$\left(\frac{\partial \dot{\bar{\varphi}}}{\partial \dot{\mathbf{q}}}\right) = \left(\frac{\partial \bar{\varphi}}{\partial \mathbf{q}}\right). \tag{1.115}$$

With the parametrization $\bar{\varphi} = \bar{\varphi}(\mathbf{q},t)$, it is $\dot{\bar{\varphi}} = \dot{\bar{\varphi}}(\mathbf{q},\dot{\mathbf{q}},t)$ and $\omega = \omega(\mathbf{q},\dot{\mathbf{q}},t)$. It follows

$$\mathbf{J}_R = \mathbf{Y}^{-1}(\bar{\varphi}(\mathbf{q},t))\left(\frac{\partial \dot{\bar{\varphi}}}{\partial \dot{\mathbf{q}}}\right). \tag{1.116}$$

The relation

$$\mathbf{J}_R = \left(\frac{\partial \dot{\bar{\varphi}}}{\partial \dot{\mathbf{q}}}\right) \tag{1.117}$$

can be used for plane rotations only.

Example 1.7 (Car with a rotating disc). To illustrate the moment of momentum equation, we consider an automobile with remote control and a high-speed gyro [29]. Figure 1.17 depicts the operation. The car drives along a left-hand bend, which has the effect of a constraint torque $\mathbf{M}_S(t)$. Applying the moment of momentum equation in a very simple form

$$\Delta\mathbf{L}_S(t) = \mathbf{M}_S(t)\Delta t,$$

we state the following:

- On the left side of Fig. 1.17, the moment of momentum vector $\mathbf{L}_S(t)$ of the high-speed gyro points towards the left side. It will elevated by $\Delta\mathbf{L}_S(t)$. Consequently, the inside wheels lift off the ground.
- On the right side of Fig. 1.17, the moment of momentum vector $\mathbf{L}_S(t)$ of the high-speed gyro points towards the right side. It will elevated by $\Delta\mathbf{L}_S(t)$. Consequently, the outside wheels lift off the ground.

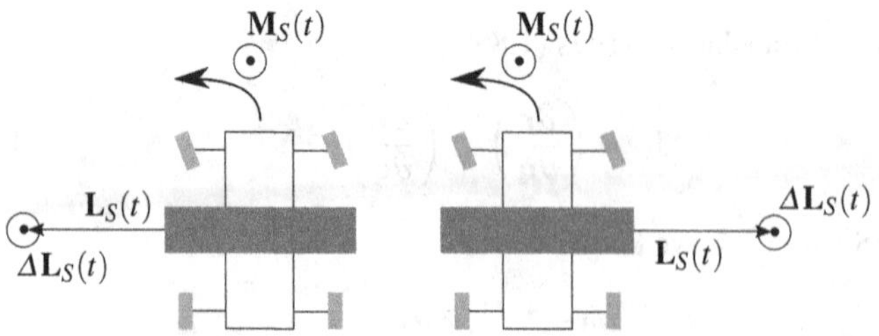

Fig. 1.17. Car with remote control – moment of momentum.

Example 1.8 (Edge-runner mill). Edge-runner mills have been used for thousands of years, but they are still a very important machine component for example in chemical engineering plants. Two wheels are connected by a rigid axis and are driven by

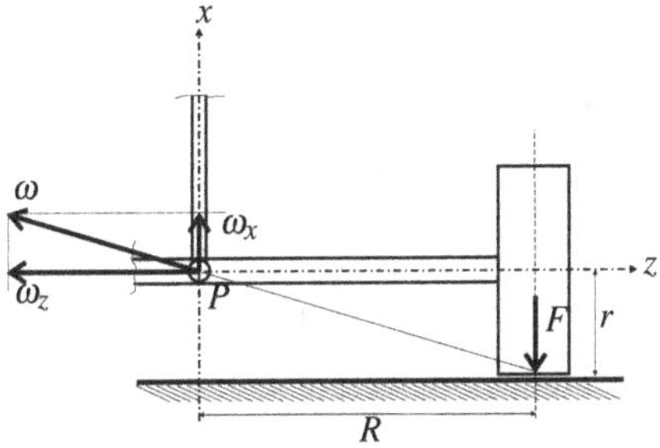

Fig. 1.18. Edge-runner mill.

a vertical axis (Fig. 1.18). It makes sense for this case to introduce an intermediate coordinate system, which is neither body-fixed nor inertial, and which rotates with the angular vertical velocity component ω_x but not with the horizontal angular velocity component ω_z. Nevertheless, it is no inertial system and therefore (1.95) applies. We write

$$\dot{\mathbf{L}}_P + \omega \times \mathbf{L}_P = \mathbf{M}_P, \qquad \text{with} \quad \omega = (\omega_x, 0, 0)^T, \quad \mathbf{L}_P = (A\omega_x, 0, C\omega_z)^T,$$

where A and C are the moments of inertia for the x- and z-directions, respectively. For a stationary rotation, we have $\dot{\omega}_x = 0, \dot{\omega}_z = 0$ and therefore also $\dot{\mathbf{L}}_P = \mathbf{0}$. The gyroscopic torque follows with

$$\mathbf{M}_P = \omega \times \mathbf{L}_P = (0, -C\omega_x\omega_z, 0)^T = (M_x, M_y, M_z)^T.$$

This result may be used to calculate the pressure acting on the ground by the rotating wheels. Taking into account the rolling condition $(r\omega_z + R\omega_x = 0)$, we obtain

$$F = \frac{|M_y|}{R} = (\frac{C}{r})\omega_x^2.$$

Example 1.9 (Airplane in a loop). An airplane with one powertrain only is driven by a propeller-engine unit. It has a moment of momentum $\mathbf{L}$ (Fig. 1.19). If such an airplane is starting, landing, or performing a loop, this moment of momentum vector is rotated around the y-axis generating a small moment of momentum vector $\Delta\mathbf{L}$ in the z-direction. This change of moment of momentum gives a yaw to the airplane around the z-axis, which of course should not happen. Therefore, this additional moment of momentum is usually counterbalanced by a trimming tab.

The relevant vectors are the following:

$$\begin{aligned} \omega_P &= (\omega_x, 0, 0)^T && \text{propeller and engine},\\ \omega_S &= (0, \omega_y, 0)^T && \text{start, landing, loops},\\ \mathbf{L} &= (L_x, 0, 0)^T && \text{moment of momentum of propeller and engine},\\ \Delta\mathbf{L} &= (0, 0, \Delta L_z)^T && \text{moment of momentum change, yaw around } z\text{-axis}. \end{aligned}$$

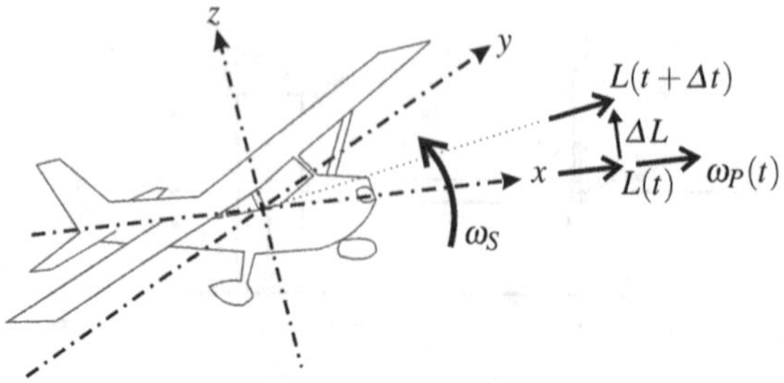

Fig. 1.19. Starting aircraft.

1.9 Lagrange's Equations

In the Sections 1.7 and 1.8, we presented some direct methods for interconnected mechanical systems, where interconnection is described by constraints of a certain kinematic-algebraic form. The dynamics of such systems has been evaluated by projecting the motion into the not-constrained directions. In special cases where the equations of motion can be formulated by minimal coordinates, we satisfy automatically all constraints and then we apply any solution procedure. Otherwise we must stay with the projection equations together with the constraints, that is differential algebraic equations (DAE), which require special treatment. In the following, we consider these procedures allowing a determination of motion in the presence of constraints [5, 63, 11, 51].

1.9.1 Lagrange's Equations of the First Kind

We consider a system of n interconnected rigid bodies and describe the velocity of a single rigid body i by (1.82)

$$\dot{\mathbf{r}}_i = (\mathbf{v}_{O'} + \tilde{\omega}\mathbf{r}_{O'P})_i = \left(\mathbf{v}_{O'} + \tilde{\mathbf{r}}_{O'P}^T\omega\right)_i. \tag{1.118}$$

Then, we choose a set of velocity coordinates

$$\dot{\mathbf{z}}_i^T = \left(\mathbf{v}_{O'}^T \;\; \omega^T\right)_i \in \mathrm{IR}^6 \tag{1.119}$$

and determine the virtual velocity vector

$$\delta\dot{\mathbf{r}}_i = \left(\mathbf{I} \;\; \tilde{\mathbf{r}}_{O'P}^T\right)_i \begin{pmatrix} \delta\mathbf{v}_{O'} \\ \delta\omega \end{pmatrix}_i. \tag{1.120}$$

The accelerations follow from the relations (1.83). The virtual power of the complete interconnected system is given by the arguments connected with JOURDAIN's principle (Section 1.7)

$$\sum_{i=1}^{n} \int_{K_i} \delta \dot{\mathbf{r}}_i^T \left(\ddot{\mathbf{r}}_i \mathrm{d}m_i - \mathrm{d}\mathbf{F}_i^e - \mathrm{d}\mathbf{F}_i^z \right) = 0 \tag{1.121}$$

and further on by (1.83) and (1.120)

$$\sum_{i=1}^{n} \int_{K_i} \begin{pmatrix} \delta \mathbf{v}_{O'} \\ \delta \omega \end{pmatrix}_i^T \begin{pmatrix} \mathbf{I} \\ \tilde{\mathbf{r}}_{O'P} \end{pmatrix}_i \left\{ \left[\mathbf{a}_{O'} + \left(\dot{\tilde{\omega}} + \tilde{\omega}\tilde{\omega} \right) \mathbf{r}_{O'P} \right] \mathrm{d}m - \mathrm{d}\mathbf{F}^e - \mathrm{d}\mathbf{F}^z \right\}_i = 0. \tag{1.122}$$

Considering also (1.95), we obtain

$$\sum_{i=1}^{n} \begin{pmatrix} \delta \mathbf{v}_{O'} \\ \delta \omega \end{pmatrix}_i^T \begin{pmatrix} m\mathbf{a}_{O'} + m\tilde{\mathbf{r}}_{O'S}^T \dot{\omega} + m\tilde{\omega}\tilde{\omega}\mathbf{r}_{O'S} - \mathbf{F}^e - \mathbf{F}^z \\ m\tilde{\mathbf{r}}_{O'S}\mathbf{a}_{O'} + \Theta_{O'}\dot{\omega} + \tilde{\omega}\Theta_{O'}\omega - \mathbf{M}_{O'}^e - \mathbf{M}_{O'}^z \end{pmatrix}_i = 0 \tag{1.123}$$

or with better organization

$$\sum_{i=1}^{n} \delta \underbrace{\begin{pmatrix} \mathbf{v}_{O'} \\ \omega \end{pmatrix}}_{\dot{\mathbf{z}}}{}_i^T \left\{ \underbrace{\begin{pmatrix} m\mathbf{I} & m\tilde{\mathbf{r}}_{O'S}^T \\ m\tilde{\mathbf{r}}_{O'S} & \Theta_{O'} \end{pmatrix}}_{\mathbf{M}} \underbrace{\begin{pmatrix} \mathbf{a}_{O'} \\ \dot{\omega} \end{pmatrix}}_{\ddot{\mathbf{z}}} + \underbrace{\begin{pmatrix} m\tilde{\omega}\tilde{\omega}\mathbf{r}_{O'S} \\ \tilde{\omega}\Theta_{O'}\omega \end{pmatrix}}_{\mathbf{f}^g} - \underbrace{\begin{pmatrix} \mathbf{F}^e \\ \mathbf{M}_{O'}^e \end{pmatrix}}_{\mathbf{f}^e} - \underbrace{\begin{pmatrix} \mathbf{F}^z \\ \mathbf{M}_{O'}^z \end{pmatrix}}_{\mathbf{f}^z} \right\}_i = 0. \tag{1.124}$$

All abbreviations in (1.124) are defined in an $\mathbb{R}^6$-space. With

$$\mathbf{z} := \begin{pmatrix} \mathbf{z}_1 \\ \vdots \\ \mathbf{z}_n \end{pmatrix}, \; \mathbf{f}^g := \begin{pmatrix} \mathbf{f}_1^g \\ \vdots \\ \mathbf{f}_n^g \end{pmatrix}, \; \mathbf{f}^e := \begin{pmatrix} \mathbf{f}_1^e \\ \vdots \\ \mathbf{f}_n^e \end{pmatrix}, \; \mathbf{f}^z := \begin{pmatrix} \mathbf{f}_1^z \\ \vdots \\ \mathbf{f}_n^z \end{pmatrix} \tag{1.125}$$

we arrive at the corresponding vectors in the $\mathbb{R}^{6n}$-space, and with

$$\mathbf{M} := \mathrm{diag}\left(\mathbf{M}_i\right) \in \mathbb{R}^{6n,6n} \tag{1.126}$$

we define the *mass matrix* of the total system. From this, we write

$$\delta \dot{\mathbf{z}}^T \left(\mathbf{M}\ddot{\mathbf{z}} + \mathbf{f}^g - \mathbf{f}^e - \mathbf{f}^z \right) = 0. \tag{1.127}$$

The virtual velocities $\delta \dot{\mathbf{z}}$ cannot be chosen arbitrarily, but they must satisfy the constraints (1.78), namely

$$\Phi\left(\mathbf{z}, \dot{\mathbf{z}}, t\right) = \mathbf{W}\left(\mathbf{z}, t\right)^T \dot{\mathbf{z}} + \hat{\mathbf{w}}\left(\mathbf{z}, t\right) = \mathbf{0} \quad \text{with} \quad \dot{\mathbf{z}} \in \mathbb{R}^{6n}, \quad \Phi \in \mathbb{R}^m, \quad m < 6n. \tag{1.128}$$

On the one hand with (1.79), we have

$$\delta \Phi = \mathbf{W}(\mathbf{z}, t)^T \delta \dot{\mathbf{z}} = \mathbf{0} \tag{1.129}$$

including the JACOBIAN *of the generalized force directions* $\mathbf{W} \in \mathrm{I\!R}^{6n,m}$. Equation (1.129) confirms the fact that the columns of $\mathbf{W}(\mathbf{z},t)$ are perpendicular to $\delta\dot{\mathbf{z}}$.

On the other hand, the principle of JOURDAIN (1.77) writes

$$0 = \sum_{i=1}^{n} \int_{K_i} \delta\dot{\mathbf{r}}_i^T \mathrm{d}\mathbf{F}_i^z = \sum_{i=1}^{n} \int_{K_i} \begin{pmatrix} \delta\mathbf{v}_{O'} \\ \delta\omega \end{pmatrix}_i^T \begin{pmatrix} \mathbf{I} \\ \tilde{\mathbf{r}}_{O'P} \end{pmatrix}_i \mathrm{d}\mathbf{F}_i^z = \delta\dot{\mathbf{z}}^T \mathbf{f}^z, \tag{1.130}$$

which says that the vector of the generalized constraint forces is perpendicular also to the virtual velocity $\delta\dot{\mathbf{z}}$. This allows us to represent the constraint forces as a linear combination of the columns of $\mathbf{W}$ giving

$$\mathbf{f}^z = -\mathbf{W}(\mathbf{z},t)\lambda \quad \text{with } \lambda \in \mathrm{I\!R}^m. \tag{1.131}$$

We combine the two relations (1.131) and (1.127) thus taking into consideration the constraints. Consequently, the virtual velocities may take any values compatible with the constraints. Furthermore, applying the fundamental lemma of variational calculus we finally get $(6n+m)$ linear equations for the unknown quantities $\ddot{\mathbf{z}} \in \mathrm{I\!R}^{6n}$ and $\lambda \in \mathrm{I\!R}^m$

$$\begin{pmatrix} \mathbf{M} & \mathbf{W} \\ \mathbf{W}^T & \mathbf{0} \end{pmatrix} \begin{pmatrix} \ddot{\mathbf{z}} \\ \lambda \end{pmatrix} + \begin{pmatrix} \mathbf{f}^g - \mathbf{f}^e \\ \bar{\mathbf{w}} \end{pmatrix} = \begin{pmatrix} \mathbf{0} \\ \mathbf{0} \end{pmatrix}. \tag{1.132}$$

Instead of the original constraint, we use the corresponding (hidden) constraint on the acceleration level (Section 1.3.3) for the above equation. Equation (1.132) represents a *saddle point* relation, because the solution of

$$\begin{pmatrix} \mathbf{M} & \mathbf{W} \\ \mathbf{W}^T & \mathbf{0} \end{pmatrix} \begin{pmatrix} \ddot{\mathbf{z}} \\ \lambda \end{pmatrix} = \begin{pmatrix} \mathbf{0} \\ \mathbf{0} \end{pmatrix} \tag{1.133}$$

is a saddle point of the quadratic form

$$\begin{pmatrix} \ddot{\mathbf{z}}^T & \lambda^T \end{pmatrix} \begin{pmatrix} \mathbf{M} & \mathbf{W} \\ \mathbf{W}^T & \mathbf{0} \end{pmatrix} \begin{pmatrix} \ddot{\mathbf{z}} \\ \lambda \end{pmatrix}. \tag{1.134}$$

Solving the first set of equations in (1.132) for $\ddot{\mathbf{z}}$

$$\ddot{\mathbf{z}} = -\mathbf{M}^{-1}(\mathbf{f}^g - \mathbf{f}^e + \mathbf{W}\lambda) \tag{1.135}$$

and including that into the second set of equations in (1.132) results in an expression for the LAGRANGE-multiplier

$$\lambda = -(\mathbf{W}^T\mathbf{M}^{-1}\mathbf{W})^{-1}\left[\mathbf{W}^T\mathbf{M}^{-1}(\mathbf{f}^g - \mathbf{f}^e) - \bar{\mathbf{w}}\right]. \quad (1.136)$$

Inserting (1.136) in (1.135) gives a reduced set of differential equations for $\mathbf{z}$, which satisfies all constraints

Example 1.10 (Null space matrix for a pendulum). We once again consider the pendulum of Fig. 1.2 with $\mathbf{z} = \mathbf{r}$ and introduce the minimal coordinate $q = \vartheta$ (Fig. 1.20). The Cartesian position $\mathbf{r}$ can be parameterized by the minimal coordinate q:

$$\mathbf{r} = \mathbf{r}(q) = \begin{pmatrix} R\sin\vartheta & -R\cos\vartheta & 0 \end{pmatrix}^T.$$

Using the coordinates $\mathbf{r}$, we have

$$m\ddot{\mathbf{r}} = m\mathbf{g} + \mathbf{F}^z,$$
$$\Phi(\mathbf{r}(q)) = \mathbf{r}(q)^T\mathbf{r}(q) - R^2 = 0,$$

where the holonomic-scleronomic constraints are satisfied automatically [21, 28, 41]. The JACOBIAN with respect to the mass center

$$\mathbf{J}_S = \frac{\partial \mathbf{r}}{\partial q} = \begin{pmatrix} R\cos\vartheta & R\sin\vartheta & 0 \end{pmatrix}^T$$

gives the actual free motion direction of the point mass. We get

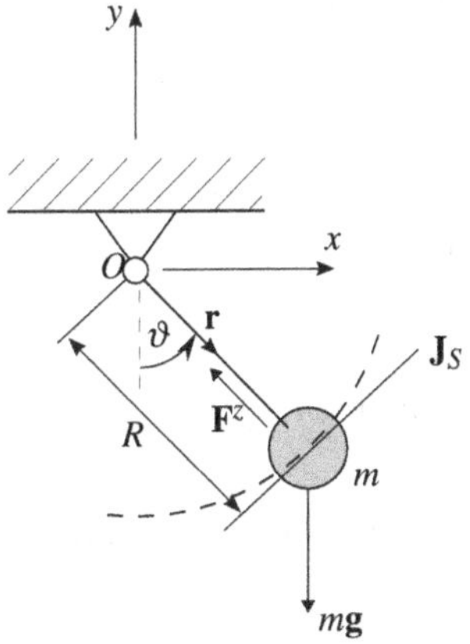

Fig. 1.20. Null space matrix of a pendulum.

$$0 = \frac{\partial \Phi}{\partial q} = \frac{\partial \Phi}{\partial \mathbf{r}} \frac{\partial \mathbf{r}}{\partial q} = 2\mathbf{r}^T \mathbf{J}_S.$$

The JACOBIAN of translation represents the null space matrix due to the special selection of $\mathbf{z} = \mathbf{r}$, and it projects the applied forces (gravitation) into the free direction, namely along the circle around the suspension point of the pendulum. Only that component can influence actively the energy budget of the pendulum. The equations of motion for q are

$$m\mathbf{J}_S^T \ddot{\mathbf{r}} = m\mathbf{J}_S^T \mathbf{g} + \underbrace{\mathbf{J}_S^T \mathbf{F}^z}_{=0}.$$

Example 1.10 of a pendulum considers minimal and nonminimal parametrizations, $\mathbf{q}$ and $\mathbf{z}$ respectively, and introduces a null space matrix $\frac{\partial \mathbf{z}}{\partial \mathbf{q}}$, the rows of which are perpendicular to the columns of the JACOBIAN of the generalized force directions $\mathbf{W}$ ((1.129) and (1.130)). This more formal property means that the rows of the null space matrix must have the direction of the tangent vectors to the constraint surfaces, because the generalized constraint forces can only be part of the normal space of these surfaces.

Coming back to (1.132) and searching for some parametrization $\mathbf{z}(\mathbf{q})$ by a set of coordinates $\mathbf{q}$, we are able to satisfy the constraints automatically. Projecting the relations (1.132) into the free directions of motion by multiplying in a further step with the transposed null space matrix $(\frac{\partial \mathbf{z}}{\partial \mathbf{q}})^T$ from the left-hand side, we arrive at the NEWTON-EULER equations formulated in minimal coordinates (Section 1.8). Their solution represents a trajectory in time expressed by the minimal coordinates $\mathbf{q}$. If in addition the generalized constraint forces are needed, for design purposes for example, they might be evaluated by the relation $\mathbf{z}(\mathbf{q})$ applied to (1.136), which sometimes is called *inverse kinetics*.

1.9.2 Lagrange's Equations of the Second Kind

1.9.2.1 Derivation

We restrict our derivation to holonomic constraints, more general constraints follow however similar ideas as presented here [28, 50]. Starting with the principle of d'ALEMBERT (1.71) and assuming a multibody system, we write

$$\sum_{i=1}^{n} \int_{K_i} (\ddot{\mathbf{r}} dm - d\mathbf{F}^e)_i^T \, \delta \mathbf{r}_i = 0. \tag{1.137}$$

To carry out some manipulation with this expression, we consider the kinetic energy of all single bodies

$$T = \sum_{i=1}^{n} T_i = \sum_{i=1}^{n} \frac{1}{2} \int_{K_i} \dot{\mathbf{r}}_i^T \dot{\mathbf{r}}_i dm_i. \tag{1.138}$$

For the following calculations, we consider a single rigid body and combine later the results for a multibody system. We modify the acceleration in (1.137) to the following form

$$\int_{K_i} \ddot{\mathbf{r}}_i^T \mathrm{d}m_i \delta \mathbf{r}_i = \frac{\mathrm{d}}{\mathrm{d}t} \int_{K_i} \dot{\mathbf{r}}_i^T \mathrm{d}m_i \delta \mathbf{r}_i - \int_{K_i} \dot{\mathbf{r}}_i^T \mathrm{d}m_i \delta \dot{\mathbf{r}}_i$$
$$= \frac{\mathrm{d}}{\mathrm{d}t} \int_{K_i} \frac{\partial}{\partial \dot{\mathbf{r}}_i} \left(\frac{1}{2} \dot{\mathbf{r}}_i^T \dot{\mathbf{r}}_i \mathrm{d}m_i \right) \delta \mathbf{r}_i - \delta \int_{K_i} \left(\frac{1}{2} \dot{\mathbf{r}}_i^T \dot{\mathbf{r}}_i \mathrm{d}m_i \right). \tag{1.139}$$

This reshaping uses the exchangeability of integration, differentiation, and variation due to the linear properties of the equations and due to constant areas of integration [20, 28]. With the minimal coordinates $\mathbf{r}_i = \mathbf{r}_i(\mathbf{q}, t)$ ($\mathbf{r}_i \in \mathrm{I\!R}^3$, $\mathbf{q} \in \mathrm{I\!R}^f$), (1.85), and variation with respect to $\delta \mathbf{q}$, we get ($\delta t = 0$):

$$\delta \mathbf{r}_i = \left(\frac{\partial \mathbf{r}_i}{\partial \mathbf{q}} \right) \delta \mathbf{q} = \left(\frac{\partial \dot{\mathbf{r}}_i}{\partial \dot{\mathbf{q}}} \right) \delta \mathbf{q}. \tag{1.140}$$

Together with (1.138), the relation (1.139) gives

$$\int_{K_i} \ddot{\mathbf{r}}_i^T \mathrm{d}m_i \delta \mathbf{r}_i = \frac{\mathrm{d}}{\mathrm{d}t} \int_{K_i} \frac{\partial}{\partial \dot{\mathbf{r}}_i} \left(\frac{1}{2} \dot{\mathbf{r}}_i^T \dot{\mathbf{r}}_i \mathrm{d}m_i \right) \left(\frac{\partial \dot{\mathbf{r}}_i}{\partial \dot{\mathbf{q}}} \right) \delta \mathbf{q} - \delta T_i$$
$$= \frac{\mathrm{d}}{\mathrm{d}t} \int_{K_i} \frac{\partial}{\partial \dot{\mathbf{q}}} \left(\frac{1}{2} \dot{\mathbf{r}}_i^T \dot{\mathbf{r}}_i \mathrm{d}m_i \right) \delta \mathbf{q} - \delta T_i$$
$$= \frac{\mathrm{d}}{\mathrm{d}t} \left(\frac{\partial T_i}{\partial \dot{\mathbf{q}}} \delta \mathbf{q} \right) - \delta T_i. \tag{1.141}$$

We vary the kinetic energy $T_i = T_i(\mathbf{q}, \dot{\mathbf{q}}, t)$ without considering time ($\delta t = 0$, see Section 1.3.4)

$$\delta T_i = \frac{\partial T_i}{\partial \mathbf{q}} \delta \mathbf{q} + \frac{\partial T_i}{\partial \dot{\mathbf{q}}} \delta \dot{\mathbf{q}} \tag{1.142}$$

and get

$$\int_{K_i} \ddot{\mathbf{r}}_i^T \mathrm{d}m_i \delta \mathbf{r}_i = \left[\frac{\mathrm{d}}{\mathrm{d}t} \left(\frac{\partial T_i}{\partial \dot{\mathbf{q}}} \right) - \frac{\partial T_i}{\partial \mathbf{q}} \right] \delta \mathbf{q} + \frac{\partial T_i}{\partial \dot{\mathbf{q}}} \left[\frac{d}{\mathrm{d}t} (\delta \mathbf{q}) - \delta \dot{\mathbf{q}} \right]. \tag{1.143}$$

Again exchanging differentiation and variation, the last term in (1.143) vanishes ($\frac{d}{\mathrm{d}t} (\delta \mathbf{q}) - \delta \dot{\mathbf{q}} = \mathbf{0}$) [28, 50]. The virtual work of the applied forces $\mathrm{d}\mathbf{F}_i^e$ is

$$\delta W_i^e = \int_{K_i} (\mathrm{d}\mathbf{F}_i^e)^T \delta \mathbf{r}_i = \int_{K_i} (\mathrm{d}\mathbf{F}_i^e)^T \left(\frac{\partial \mathbf{r}_i}{\partial \mathbf{q}} \right) \delta \mathbf{q} = \mathbf{Q}_i^T \delta \mathbf{q} \tag{1.144}$$

where

$$\mathbf{Q}_i = \int_{K_i} \left(\frac{\partial \mathbf{r}_i}{\partial \mathbf{q}}\right)^T \mathrm{d}\mathbf{F}_i^e = \int_{K_i} \left(\frac{\partial \dot{\mathbf{r}}_i}{\partial \dot{\mathbf{q}}}\right)^T \mathrm{d}\mathbf{F}_i^e \tag{1.145}$$

are the *generalized forces* of the body K_i. Combining (1.137), (1.143), and (1.145), we come out with

$$\sum_{i=1}^{n} \left[\frac{\mathrm{d}}{\mathrm{d}t}\left(\frac{\partial T_i}{\partial \dot{\mathbf{q}}}\right) - \left(\frac{\partial T_i}{\partial \mathbf{q}}\right) - \mathbf{Q}_i^T\right] \delta\mathbf{q} = 0. \tag{1.146}$$

With $\mathbf{q}$ being minimal coordinates and therefore $\delta\mathbf{q}$ being arbitrary, we achieve LAGRANGE's equations of motion of the second kind by the fundamental lemma of variational calculus

$$\sum_{i=1}^{n} \left[\frac{\mathrm{d}}{\mathrm{d}t}\left(\frac{\partial T_i}{\partial \dot{\mathbf{q}}}\right) - \left(\frac{\partial T_i}{\partial \mathbf{q}}\right) - \mathbf{Q}_i^T\right] = \mathbf{0}. \tag{1.147}$$

The generalized forces $\mathbf{Q}_i$ can be subdivided into conservative and nonconservative forces, where the first ones, $\mathbf{Q}_K$, can be derived from a potential V (Section 1.6):

$$\mathbf{Q}_{K_i} = -\left(\frac{\partial V_i}{\partial \mathbf{q}}\right)^T. \tag{1.148}$$

Most mechanical systems include generalized conservative forces $\mathbf{Q}_K$ as well as generalized nonconservative forces $\mathbf{Q}_{NK}$. Introducing

$$V = \sum_{i=1}^{n} V_i, \tag{1.149}$$

$$\mathbf{Q}_{NK} = \sum_{i=1}^{n} \int_{K_i} \left(\frac{\partial \mathbf{r}_i}{\partial \mathbf{q}}\right)^T \mathrm{d}\mathbf{F}_{NK_i}^e, \tag{1.150}$$

we get f LAGRANGE's equations of the second kind in the following compact form:

$$\frac{\mathrm{d}}{\mathrm{d}t}\left(\frac{\partial T}{\partial \dot{\mathbf{q}}}\right) - \left(\frac{\partial T}{\partial \mathbf{q}}\right) + \left(\frac{\partial V}{\partial \mathbf{q}}\right) = \mathbf{Q}_{NK}^T. \tag{1.151}$$

The special case of $\frac{\mathrm{d}}{\mathrm{d}t}\left(\frac{\partial T}{\partial \dot{\mathbf{q}}_s}\right) = 0$ (cyclic momenta) will be treated in Section 4.4.3 and in the following examples.

1.9.2.2 Remarks on Evaluation

Lagrange's equations (1.151) have been derived with respect to arbitrary rigid mechanical bodies. Their derivation by the virtual work principle and their formulation in terms of energy even allows us to apply them to physical systems of a completely different nature [25]. We restrict our considerations to rigid bodies.

The kinetic energy of a single body writes by combination of (1.138), (1.82), and (1.93):

$$\begin{aligned} T_i &= \frac{1}{2}\int_{K_i} [\mathbf{v}_{O'} + \tilde{\omega}\mathbf{r}_{O'P}]_i^T [\mathbf{v}_{O'} + \tilde{\omega}\mathbf{r}_{O'P}]_i \,\mathrm{d}m_i \\ &= \left\{\frac{1}{2}m\mathbf{v}_{O'}^T\mathbf{v}_{O'} + m\mathbf{v}_{O'}^T\tilde{\omega}\mathbf{r}_{O'S} + \frac{1}{2}\int_{K_i}[-\tilde{\mathbf{r}}_{O'P}\omega]^T[-\tilde{\mathbf{r}}_{O'P}\omega]\,\mathrm{d}m\right\}_i \\ &= \left\{\frac{1}{2}m\mathbf{v}_{O'}^T\mathbf{v}_{O'} + m\mathbf{v}_{O'}^T\tilde{\omega}\mathbf{r}_{O'S} + \frac{1}{2}\omega^T\Theta_{O'}\omega\right\}_i . \end{aligned} \tag{1.152}$$

This formula has three parts: a term concerning translation, a coupling term concerning both translation and rotation, and a rotational term. The coupling term vanishes for the mass center as the reference $O'_i = S_i$ and thus $\mathbf{r}_{O'_iS_i} = \mathbf{0}$. Then, we get

$$T_i = \left\{\frac{1}{2}m\mathbf{v}_S^T\mathbf{v}_S + \frac{1}{2}\omega^T\Theta_S\omega\right\}_i . \tag{1.153}$$

The potential energy very often consists of a general spring potential [15] and a potential for gravity giving

$$V_f = \frac{1}{2}\left(\mathbf{r}_{F_1F_2} - \mathbf{r}^0_{F_1F_2}\right)^T \mathbf{C}\left(\mathbf{r}_{F_1F_2} - \mathbf{r}^0_{F_1F_2}\right), \tag{1.154}$$

$$V_g = -m\mathbf{g}^T\mathbf{r}_{OS}. \tag{1.155}$$

The vector $\mathbf{r}_{F_1F_2}$ is the distance of the spring end points, $\mathbf{r}^0_{F_1F_2}$ is the distance for the springs without load, $\mathbf{C}$ a positive-definite matrix of the spring constants, $\mathbf{g}$ the gravity acceleration, and $\mathbf{r}_{OS}$ the vector from an inertial point to the mass centers, of course evaluated in the same coordinate frame. For multibody systems another representation is frequently applied, namely

$$V_f = \frac{1}{2}c(\| \mathbf{r}_{F_1F_2} \| - l_0)^2, \tag{1.156}$$

$$V_f = \frac{1}{2}c(\alpha - \alpha_0)^2 \tag{1.157}$$

suitable for translational springs with the spring constant c and the load-free length l_0 and for rotational springs with fixed axis, a turning angle α, and a load-free angular displacement of α_0.

Nonconservative forces are related to their point of application using (1.85).

$$\mathbf{Q}_{NK} = \sum_{i=1}^{n}\int_{K_i}\left(\frac{\partial \mathbf{r}_i}{\partial \mathbf{q}}\right)^T \mathrm{d}\mathbf{F}^e_{NK_i} = \sum_{i=1}^{n}\int_{K_i}\left(\frac{\partial \dot{\mathbf{r}}_i}{\partial \dot{\mathbf{q}}}\right)^T \mathrm{d}\mathbf{F}^e_{NK_i} \tag{1.158}$$

For a rigid body we apply (1.82) and use as a reference point O' regarding in addition the definitions of Section 1.8. It is

$$\mathbf{Q}_{NK} = \sum_{i=1}^{n} \int_{K_i} \left[\left(\frac{\partial \mathbf{v}_{O'}}{\partial \dot{\mathbf{q}}} \right)^T + \left(\frac{\partial \omega}{\partial \dot{\mathbf{q}}} \right)^T \tilde{\mathbf{r}}_{O'P} \right]_i \mathrm{d}\mathbf{F}^e_{NK_i}$$
$$= \sum_{i=1}^{n} \left(\mathbf{J}^T_{O'_i} \mathbf{F}^e_{NK_i} + \mathbf{J}^T_{R_i} \mathbf{M}^e_{NK_i,O'} \right). \quad (1.159)$$

This procedure introduces a torque

$$\mathbf{M}^e_{NK_i,O'} = \int_{K_i} [\tilde{\mathbf{r}}_{O'P}]_i \mathrm{d}\mathbf{F}^e_{NK_i} \quad (1.160)$$

meaning that applied free torques can also be projected by the JACOBIANs of rotation into the free directions of motion. For forces, one uses directly, if possible, the JACOBIANs of translation with respect to the points of force applications.

Lagrange's equations of the second kind offer an analytical route to the equations of motion, provided, we are able to find a set of minimal coordinates. For many practical cases this will be possible, and then we proceed stepwise in the following form:

- Look for a convenient set of minimal coordinates $\mathbf{q} \in \mathbb{R}^f$ with respect to the problem,
- evaluate the kinetic energies T_i and the potential energies V_i,
- calculate the generalized nonconservative forces and torques $\mathbf{Q}_{NK}$,
- evaluate Lagrange's equations of the second kind.

1.9.2.3 Examples

Example 1.11 (KEPLER's laws of planetary trajectories). The famous three laws of Johannes KEPLER write:

1. The orbit of every planet is an ellipse with the sun at one of the two foci (ellipse law).
2. A line joining a planet and the sun sweeps out equal areas during equal intervals of time (area law).
3. The square of the orbital period of a planet is directly proportional to the cube of the semimajor axis of its orbit.

We prove the second law using LAGRANGE's equations of the second kind (Fig. 1.21). We choose minimal coordinates $\mathbf{q} = (r, \varphi)^T$ and describe the vector from the point M to the planet mass m and the angular velocity of the moving coordinate system by

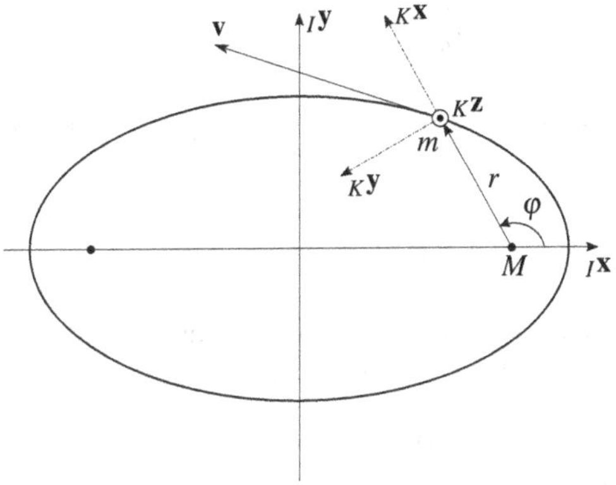

Fig. 1.21. KEPLER's laws.

$$
\begin{aligned}
{}_K\mathbf{r} &= \begin{pmatrix} r & 0 & 0 \end{pmatrix}^T, \\
{}_K\omega &= \begin{pmatrix} 0 & 0 & \dot{\varphi} \end{pmatrix}^T.
\end{aligned}
$$

The absolute velocity of m then writes

$$
\begin{aligned}
{}_K\mathbf{v} &= \begin{pmatrix} \dot{r} & 0 & 0 \end{pmatrix}^T + \begin{pmatrix} 0 & 0 & \dot{\varphi} \end{pmatrix}^T \times \begin{pmatrix} r & 0 & 0 \end{pmatrix}^T \\
&= \begin{pmatrix} \dot{r} & 0 & 0 \end{pmatrix}^T + \begin{pmatrix} 0 & r\dot{\varphi} & 0 \end{pmatrix}^T = \begin{pmatrix} 1 & 0 \\ 0 & r \\ 0 & 0 \end{pmatrix} \dot{\mathbf{q}},
\end{aligned}
$$

giving the kinetic energy as

$$
T = \frac{1}{2} m \, {}_K\mathbf{v}^T \, {}_K\mathbf{v} = \frac{1}{2} \dot{\mathbf{q}}^T \begin{pmatrix} m & 0 \\ 0 & mr^2 \end{pmatrix} \dot{\mathbf{q}}.
$$

The gravity potential of the central force field is

$$
V = \gamma \frac{Mm}{r^2}
$$

with the *gravitational constant* γ. Nonconservative forces do not exist, thus Lagrange's equations result in

$$
\begin{pmatrix} m & 0 \\ 0 & mr^2 \end{pmatrix} \ddot{\mathbf{q}} + \begin{pmatrix} -2\gamma \frac{mM}{r^3} \\ 0 \end{pmatrix} = \begin{pmatrix} 0 \\ 0 \end{pmatrix}.
$$

The second row of this equation indicates a *cyclic coordinate* accompanied by a cyclic (generalized) momentum, which is

$$p = \frac{\partial T}{\partial \dot{\varphi}} = mr^2 \dot{\varphi}$$

in the direction of φ and with the properties $\frac{\partial T}{\partial \varphi} = 0$ and $\frac{\partial V}{\partial \varphi} = 0$. From this, we get KEPLER's second law

$$\dot{A} = \frac{1}{2} r^2 \dot{\varphi} = \frac{p}{2m}.$$

Similar results can be obtained for a particle in any mechanical field of forces, for example for linear-elastic bearings $V = \frac{1}{2}cr^2$ with a spring constant c.

Example 1.12 (Spherical pendulum). We consider the spherical pendulum from Example 1.4; we also want to identify the cyclic coordinate as in Example 1.11 (Fig. 1.22). We choose the minimal coordinates $\mathbf{q} = (\psi, \vartheta)^T$. With respect to the K-system the vector from O to P writes

$${}_K\mathbf{r}_{OP} = \begin{pmatrix} 0 & 0 & -R \end{pmatrix}^T.$$

The angular velocity of the K-system can be described by EULER's sequence of rotation:

$${}_K\omega = \begin{pmatrix} \dot{\vartheta} & \dot{\psi} \sin\vartheta & \dot{\psi} \cos\vartheta \end{pmatrix}^T.$$

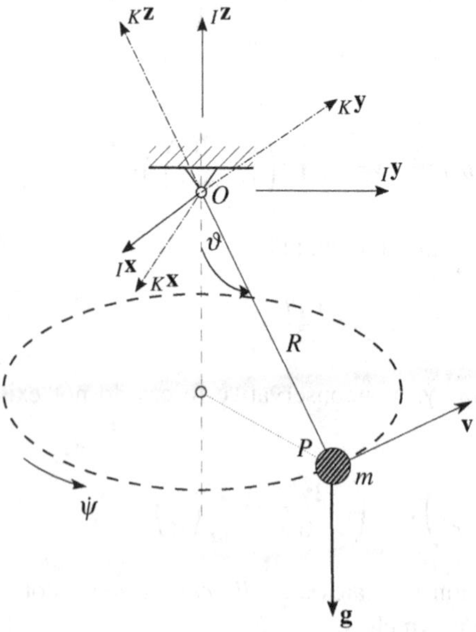

Fig. 1.22. Spherical pendulum.

With this we get the absolute velocity of the mass m with respect to the K-system

$$ {}_K\mathbf{v} = {}_K\boldsymbol{\omega} \times {}_K\mathbf{r}_{OP} = \begin{pmatrix} -R\dot{\psi}\sin\vartheta & R\dot{\vartheta} & 0 \end{pmatrix}^T $$

and finally the kinetic energy as

$$ T = \frac{1}{2} m\, {}_K\mathbf{v}^T\, {}_K\mathbf{v} = \frac{1}{2}\dot{\mathbf{q}}^T \begin{pmatrix} mR^2\sin^2\vartheta & 0 \\ 0 & mR^2 \end{pmatrix} \dot{\mathbf{q}}. $$

Introducing ${}_I\mathbf{g} = \begin{pmatrix} 0 & 0 & -g \end{pmatrix}^T$, the gravitational potential writes

$$ V = -mgR\cos\vartheta. $$

Considering all this, LAGRANGE's equations of the second kind come out with

$$ \underbrace{\begin{pmatrix} mR^2\sin^2\vartheta & 0 \\ 0 & mR^2 \end{pmatrix}\ddot{\mathbf{q}} + \begin{pmatrix} 2mR^2\sin\vartheta\cos\vartheta\,\dot{\vartheta}\,\dot{\psi} \\ 0 \end{pmatrix}}_{\frac{\mathrm{d}}{\mathrm{d}t}\left(\frac{\partial T}{\partial \dot{\mathbf{q}}}\right)} $$

$$ -\underbrace{\begin{pmatrix} 0 \\ mR^2\sin\vartheta\cos\vartheta\,\dot{\psi}\dot{\psi} \end{pmatrix}}_{\frac{\partial T}{\partial \mathbf{q}}} + \underbrace{\begin{pmatrix} 0 \\ mgR\sin\vartheta \end{pmatrix}}_{\frac{\partial V}{\partial \mathbf{q}}} = \begin{pmatrix} 0 \\ 0 \end{pmatrix}. $$

Because of $\frac{\partial T}{\partial \psi} = \frac{\partial V}{\partial \psi} = 0$, the angle ψ is a cyclic coordinate, and consequently the cyclic momentum

$$ \left(\frac{\partial T}{\partial \dot{\psi}}\right) = (mR^2\sin^2\vartheta)\dot{\psi} $$

in the direction of ψ will be constant. The projection of the string onto the horizontal plane sweeps out equal areas in equal intervals of time (Example 1.11), comparable to KEPLER's second law.

Example 1.13 (Differential gear). We consider the differential gear of Fig. 1.23 and investigate the behavior after a change in load. The differential gear consists of a drive gear 1, a hypoid gear 2, two driven gears 3,4, and two differential gears 5,6 (satellite gears) with the angular positions

$$ \mathbf{z}^T = \begin{pmatrix} \varphi_1 & \varphi_2 & \varphi_3 & \varphi_4 & \varphi_5 & \varphi_6 \end{pmatrix}^T. $$

The following kinematic conditions hold

$$ \boldsymbol{\Phi}(\mathbf{z}) = \begin{pmatrix} \varphi_1 - a_1\varphi_2 \\ \varphi_2 - \frac{1}{2}(\varphi_3 + \varphi_4) \\ \varphi_5 - a_2(\varphi_3 - \varphi_4) \\ \varphi_6 + a_2(\varphi_3 - \varphi_4) \end{pmatrix} = \mathbf{0}. $$

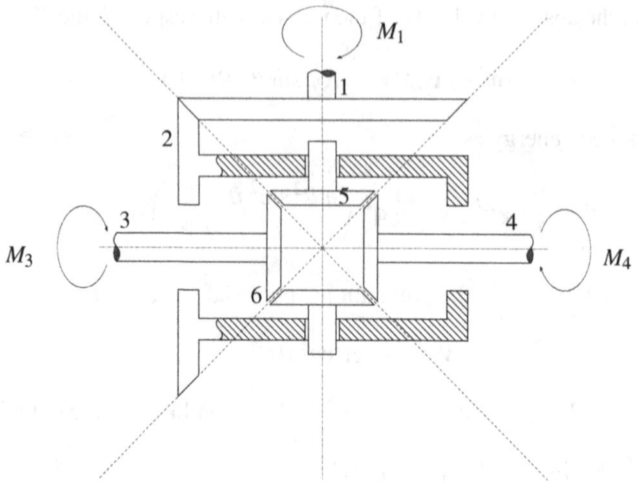

Fig. 1.23. Differential gear.

The magnitudes a_1 and a_2 are transmission ratios. The system has six bodies, four constraints, and therefore two degrees of freedom. The most convenient minimal coordinates are

$$\mathbf{q}^T = \begin{pmatrix} q_1 & q_2 \end{pmatrix}^T = \begin{pmatrix} \varphi_3 & \varphi_4 \end{pmatrix}^T .$$

They allow an easy representation of the other four coordinates

$$\mathbf{z} = \mathbf{z}(\mathbf{q}) = \begin{pmatrix} \varphi_1 \\ \varphi_2 \\ \varphi_3 \\ \varphi_4 \\ \varphi_5 \\ \varphi_6 \end{pmatrix} = \begin{pmatrix} \frac{1}{2} a_1 (q_1 + q_2) \\ \frac{1}{2}(q_1 + q_2) \\ q_1 \\ q_2 \\ a_2 (q_1 - q_2) \\ -a_2 (q_1 - q_2) \end{pmatrix} .$$

From the constraints and the last relation, we get the directions of the generalized forces in the form

$$\mathbf{W} = \begin{pmatrix} 1 & -a_1 & 0 & 0 & 0 & 0 \\ 0 & 1 & -\frac{1}{2} & -\frac{1}{2} & 0 & 0 \\ 0 & 0 & -a_2 & +a_2 & 1 & 0 \\ 0 & 0 & +a_2 & -a_2 & 0 & 1 \end{pmatrix}^T = \left(\frac{\partial \Phi}{\partial \mathbf{z}} \right)^T .$$

The JACOBIAN of $\mathbf{z}$ writes

$$\frac{\partial \mathbf{z}}{\partial \mathbf{q}} = \begin{pmatrix} \frac{a_1}{2} & \frac{a_1}{2} \\ 1/2 & 1/2 \\ 1 & 0 \\ 0 & 1 \\ a_2 & -a_2 \\ -a_2 & a_2 \end{pmatrix}.$$

With this, we project the nonconservative applied forces

$$\mathbf{f}^e = \begin{pmatrix} M_1 & 0 & -M_3 & -M_4 & 0 & 0 \end{pmatrix}^T$$

into the directions of the two minimal coordinates

$$\mathbf{Q}_{NK} = \left(\frac{\partial \mathbf{z}}{\partial \mathbf{q}}\right)^T \mathbf{f}^e = \begin{pmatrix} \frac{a_1}{2} & M_1 - M_3 \\ \frac{a_1}{2} & M_1 - M_4 \end{pmatrix}.$$

Using again $(\frac{\partial \mathbf{z}}{\partial \mathbf{q}})$ brings the kinetic energy into the form

$$T = \frac{1}{2}\dot{\mathbf{z}}^T \mathbf{M} \dot{\mathbf{z}} = \frac{1}{2}\dot{\mathbf{q}}^T \underbrace{\left[\left(\frac{\partial \mathbf{z}}{\partial \mathbf{q}}\right)^T \mathbf{M} \left(\frac{\partial \mathbf{z}}{\partial \mathbf{q}}\right)\right]}_{\mathbf{M}^* = \begin{pmatrix} M^*_{11} & M^*_{12} \\ M^*_{12} & M^*_{22} \end{pmatrix}} \dot{\mathbf{q}}$$

with $\mathbf{M} = \operatorname{diag}(J_i)$. The transverse moments of inertia of the two bodies 5,6 are included in the moment of inertia of body 2. The mass matrix projected into the direction of the two minimal coordinates is given by

$$\begin{aligned} M^*_{11} &= J_1 \left(\frac{a_1}{2}\right)^2 + \frac{1}{4}J_2 + J_3 + a_2^2 (J_5 + J_6), \\ M^*_{12} &= J_1 \left(\frac{a_1}{2}\right)^2 + \frac{1}{4}J_2 - a_2^2 (J_5 + J_6), \\ M^*_{22} &= J_1 \left(\frac{a_1}{2}\right)^2 + \frac{1}{4}J_2 + J_4 + a_2^2 (J_5 + J_6). \end{aligned}$$

Hence,

$$T = \frac{1}{2}M^*_{11}\dot{\varphi}_3^2 + M^*_{12}\dot{\varphi}_3\dot{\varphi}_4 + \frac{1}{2}M^*_{22}\dot{\varphi}_4^2.$$

If $J_3 = J_4$, then $M^*_{11} = M^*_{22}$. Generally it is $M^*_{11} > M^*_{12}$.

Evaluating LAGRANGE's equations of the second kind with $M_{11}^* = M_{22}^*$ results in

$$\left(\frac{\partial T}{\partial \dot{\varphi}_3}\right) = M_{11}^* \dot{\varphi}_3 + M_{12}^* \dot{\varphi}_4, \quad \left(\frac{\partial T}{\partial \varphi_3}\right) = 0,$$
$$\left(\frac{\partial T}{\partial \dot{\varphi}_4}\right) = M_{11}^* \dot{\varphi}_4 + M_{12}^* \dot{\varphi}_3, \quad \left(\frac{\partial T}{\partial \varphi_4}\right) = 0$$

and from this the equations of motion in minimal coordinates:

$$M_{11}^* \ddot{\varphi}_3 + M_{12}^* \ddot{\varphi}_4 = \frac{1}{2} M_1 a_1 - M_3,$$
$$M_{11}^* \ddot{\varphi}_4 + M_{12}^* \ddot{\varphi}_3 = \frac{1}{2} M_1 a_1 - M_4.$$

Usually one has to take into account the torques $M_i = M_i(\varphi_i)$ from some complicated relations that can be treated only numerically. For our considerations, we simplify that and discuss the following situations.

1. ***Stationary motion***:

$$\ddot{\varphi}_3 = \ddot{\varphi}_4 = 0$$

being possible only for $M_3 = M_4 = \frac{1}{2} M_1 a_1$.

2. ***Acceleration of the driven bodies*** (resolution for $\ddot{\varphi}_3$ and $\ddot{\varphi}_4$):

$$\ddot{\varphi}_3 = \frac{\frac{1}{2} M_1 a_1 (M_{11}^* - M_{12}^*) - M_{11}^* M_3 + M_{12}^* M_4}{M_{11}^{*2} - M_{12}^{*2}},$$
$$\ddot{\varphi}_4 = \frac{\frac{1}{2} M_1 a_1 (M_{11}^* - M_{12}^*) - M_{11}^* M_4 + M_{12}^* M_3}{M_{11}^{*2} - M_{12}^{*2}}.$$

3. ***Acceleration of the driving body***:

$$\ddot{\varphi}_1 = \frac{a_1}{2} (\ddot{\varphi}_3 + \ddot{\varphi}_4) = \frac{a_1}{2} \left(\frac{M_1 a_1 - M_3 - M_4}{M_{11}^* + M_{12}^*}\right).$$

4. ***Sudden load change***, for example ΔM_3:

$$M_3 = M_{3_0} + \Delta M_3$$

results in $\ddot{\varphi}_i = \ddot{\varphi}_{i_0} + \Delta \ddot{\varphi}_i$. Starting with a stationary motion ($\ddot{\varphi}_3 = \ddot{\varphi}_4 = 0$) after the load change, we get

$$\Delta\ddot{\varphi}_3 = -\frac{M_{11}^*\Delta M_3}{M_{11}^{*2} - M_{12}^{*2}} < 0 \qquad \text{deceleration,}$$

$$\Delta\ddot{\varphi}_4 = +\frac{M_{12}^*\Delta M_3}{M_{11}^{*2} - M_{12}^{*2}} > 0 \qquad \text{acceleration,}$$

$$\Delta\ddot{\varphi}_1 = -\frac{a_1}{2}\left(\frac{\Delta M_3}{M_{11}^{*2} - M_{12}^{*2}}\right) < 0.$$

As $M_{11}^* > M_{12}^*$ it is $\mid \Delta\ddot{\varphi}_3 \mid > \mid \Delta\ddot{\varphi}_4 \mid$.

1.10 Hamilton's Equations

HAMILTON's principle and connected with it, HAMILTON's canonical equations, were formerly of greater significance in the fields of purely theoretical mechanics but they are now moving into the fields of engineering applications due to the increasing significance of nonlinear dynamics in many fields of technology [71]. Therefore, it makes sense to discuss this approach. In the following section, we use minimal coordinates.

1.10.1 Hamilton's Principle

We start with relations (1.137), (1.141), and (1.144) from Section 1.9:

$$\sum_{i=1}^{n} \int_{K_i} (\ddot{\mathbf{r}}\mathrm{d}m - \mathrm{d}\mathbf{F}^e)_i^T \,\delta\mathbf{r}_i = 0, \tag{1.161}$$

$$\int_{K_i} \ddot{\mathbf{r}}_i^T \,\mathrm{d}m_i \delta\mathbf{r}_i = \frac{\mathrm{d}}{\mathrm{d}t}\left(\frac{\partial T_i}{\partial \dot{\mathbf{q}}}\delta\mathbf{q}\right) - \delta T_i, \tag{1.162}$$

$$\int_{K_i} (\mathrm{d}\mathbf{F}_i^e)^T \,\delta\mathbf{r}_i = \delta W_i^e. \tag{1.163}$$

Additionally, we consider the total kinetic energy, the total virtual work of the applied forces

$$T = \sum_{i=1}^{n} T_i, \tag{1.164}$$

$$\delta W^e = \sum_{i=1}^{n} \delta W_i^e \tag{1.165}$$

and the generalized momentum as the derivation of the kinetic energy with respect to the generalized minimal velocities (Example 1.11):

$$\mathbf{p}^T = \left(\frac{\partial T}{\partial \dot{\mathbf{q}}}\right). \tag{1.166}$$

These steps result in the central equation of LAGRANGE ("Zentralgleichung") [50]

$$\frac{\mathrm{d}}{\mathrm{d}t}\left(\mathbf{p}^T\,\delta\mathbf{q}\right) - \delta T - \delta W^e = 0. \tag{1.167}$$

Integrating this relation from t_1 to t_2 gives

$$\int_{t_1}^{t_2} \delta\left(T + W^e\right)\mathrm{d}t = \left(\mathbf{p}^T\,\delta\mathbf{q}\right)\Big|_{t_1}^{t_2}. \tag{1.168}$$

Assuming for the virtual displacements $\delta\mathbf{q}(t_1) = \mathbf{0}$ and $\delta\mathbf{q}(t_2) = \mathbf{0}$, we come to the form

$$\int_{t_1}^{t_2} \left(\delta T + \delta W^e\right)\mathrm{d}t = 0. \tag{1.169}$$

For conservative systems, the applied forces might be derived from potentials (Section 1.6):

$$(\mathrm{d}\mathbf{F}^e)_i^T = -\frac{\partial\,(\mathrm{d}V_i)}{\partial\mathbf{r}_i}, \tag{1.170}$$

and therefore

$$\delta W_i^e = \int_{K_i} (\mathrm{d}\mathbf{F}_i^e)^T\,\delta\mathbf{r}_i = -\delta V_i. \tag{1.171}$$

With

$$\delta V = \sum_{i=1}^{n} \delta V_i, \tag{1.172}$$

it follows

$$\int_{t_1}^{t_2} \delta\left(T - V\right)\mathrm{d}t =: \int_{t_1}^{t_2} \delta L = 0 \tag{1.173}$$

with the LAGRANGE *function* $L = T - V$. The variations δL must be compatible with the constraints. Assuming holonomic constraints, as we have done for D'ALEMBERT's principle and which allow the exchangeability of integration and variation, we finally get

HAMILTON's principle as a *variational problem* [28]:

$$\delta \int_{t_1}^{t_2} L \mathrm{d}t = 0 \quad \text{or} \quad \int_{t_1}^{t_2} L \mathrm{d}t \rightarrow \text{stationary.} \tag{1.174}$$

HAMILTON's principle is an *integral principle* which alters trajectory elements, whereas D'ALEMBERT's principle represents a *differential principle* that compares neighbouring states. The varied magnitudes of the integral principles possess the dimension of an *action* (= energy · time). Thus, the integral principles are also called *principles of least action* [22]. To solve (1.174), we consider a bundle of functions

$$\mathbf{q}_{\varepsilon}(t) = \mathbf{q}_0(t) + \varepsilon \eta_{\mathbf{q}}(t) \tag{1.175}$$

with the parameter ε and take into account the compatibility condition

$$\eta_{\mathbf{q}}(t_1) = \eta_{\mathbf{q}}(t_2) = \mathbf{0}. \tag{1.176}$$

The relation (1.174) means

$$\begin{aligned} 0 &= \frac{\mathrm{d}}{\mathrm{d}\varepsilon}\left(\int_{t_1}^{t_2} L \mathrm{d}t\right)\bigg|_{\varepsilon=0} = \int_{t_1}^{t_2}\left(\frac{\partial L}{\partial \mathbf{q}}\eta_{\mathbf{q}}(t) + \frac{\partial L}{\partial \dot{\mathbf{q}}}\dot{\eta}_{\mathbf{q}}(t)\right)\mathrm{d}t \\ &= \int_{t_1}^{t_2}\left(\frac{\partial L}{\partial \mathbf{q}}\eta_{\mathbf{q}}(t) - \frac{\mathrm{d}}{\mathrm{d}t}\left(\frac{\partial L}{\partial \dot{\mathbf{q}}}\right)\eta_{\mathbf{q}}(t)\right)\mathrm{d}t + \underbrace{\left[\frac{\partial L}{\partial \dot{\mathbf{q}}}\eta_{\mathbf{q}}(t)\right]_{t_1}^{t_2}}_{=0}. \end{aligned} \tag{1.177}$$

With $\eta_{\mathbf{q}}$ being arbitrary, we come out with the EULER-LAGRANGE *equations*

$$\frac{\mathrm{d}}{\mathrm{d}t}\left(\frac{\partial L}{\partial \dot{\mathbf{q}}}\right) - \frac{\partial L}{\partial \mathbf{q}} = \mathbf{0}^T, \tag{1.178}$$

which are LAGRANGE's equations of the second kind with $L = T - V$ (1.151).

1.10.2 Hamilton's Canonical Equations

To derive HAMILTON's canonical equations, we represent the mechanical system by minimal coordinates $\mathbf{q}$ and by generalized momentum coordinates $\mathbf{p}$ instead of the usually considered velocity coordinates $\dot{\mathbf{q}}$. We start with LAGRANGE's equations of the second kind and restrict our considerations to conservative, holonomic-scleronomic systems. More general presentations may be found in [28].

LAGRANGE's equations of the second kind (1.178) together with (1.166) result in

$$\frac{\partial L}{\partial \mathbf{q}} = \dot{\mathbf{p}}^T. \tag{1.179}$$

The LAGRANGE function depends on $\mathbf{q}$ and $\dot{\mathbf{q}}$ giving

$$\delta L = \left(\frac{\partial L}{\partial \mathbf{q}}\right) \delta \mathbf{q} + \left(\frac{\partial L}{\partial \dot{\mathbf{q}}}\right) \delta \dot{\mathbf{q}} = \dot{\mathbf{p}}^T \delta \mathbf{q} + \mathbf{p}^T \delta \dot{\mathbf{q}}, \tag{1.180}$$

on the one hand. On the other hand, the virtual change of the product $(\mathbf{p}^T \dot{\mathbf{q}})$ results in

$$\delta\left(\mathbf{p}^T \dot{\mathbf{q}}\right) = \mathbf{p}^T \delta \dot{\mathbf{q}} + \dot{\mathbf{q}}^T \delta \mathbf{p}. \tag{1.181}$$

At this point, we introduce the HAMILTON *function*

$$H(\mathbf{q}, \mathbf{p}) = \mathbf{p}^T \dot{\mathbf{q}} - L, \tag{1.182}$$

determine formally its variation

$$\delta H = \left(\frac{\partial H}{\partial \mathbf{q}}\right) \delta \mathbf{q} + \left(\frac{\partial H}{\partial \mathbf{p}}\right) \delta \mathbf{p} \tag{1.183}$$

and compare this relation with the difference between (1.181) and (1.180)

$$\delta\left(\dot{\mathbf{p}}^T \dot{\mathbf{q}} - L\right) = \dot{\mathbf{q}}^T \delta \mathbf{p} - \dot{\mathbf{p}}^T \delta \mathbf{q}, \tag{1.184}$$

then finally coming to the *canonical equations of motion* of

William HAMILTON (1805-1865):

$$\dot{\mathbf{q}}^T = \frac{\partial H}{\partial \mathbf{p}}, \quad \dot{\mathbf{p}}^T = -\frac{\partial H}{\partial \mathbf{q}}, \tag{1.185}$$

which are $2f$ differential equations of first order, the so-called HAMILTONIAN system.

The HAMILTON function H possesses a clear interpretation. We obtain from (1.182) with (1.166) and

$$T = \frac{1}{2} \dot{\mathbf{q}}^T \mathbf{M} \dot{\mathbf{q}}$$

the following result

$$H = \mathbf{p}^T \dot{\mathbf{q}} - L = \dot{\mathbf{q}}^T \mathbf{M} \dot{\mathbf{q}} - (T - V) = 2T - (T - V) = T + V.$$

The HAMILTON function simply is the sum of kinetic and potential energies, which means the total energy of a mechanical system.

1.11 Practical Considerations

Whatever method we use in mechanics, we always end up with a set of nonlinear differential equations of motion of first or second order, which are linear with respect to the accelerations but nonlinear with respect to velocities, positions, and orientations. These equations represent the *mathematical model* of the problem and are an intermediate activity of our sequence (real-world problem - mechanical modeling - mathematical modeling - numerical modeling - simulation). It should be kept in mind that mathematics can give only these results and information from a basis, which we have defined as both assumptions and constraints of the mechanical model. Therefore, establishing the mechanical model requires extreme care, empirical knowledge, and instinctive feeling combined with a good understanding of the mechanical problems, at least from a qualitative point of view. With respect to this step, large expenditure can be generated but also omitted, if done intelligently.

The degrees of freedom (DOF) as expressed by minimal coordinates determine the size of the mathematical previously also of the mechanical model. Additional simplifications might be feasible by linearization, by using invariants of motion like energy integrals of conservative systems, or by modifying the equations of motion, for example by transforming the differential equations. In any case, we should try to find a set of minimal coordinates and if this is not possible, we have to add the relevant constraints, but again trying to find a minimalistic formulation.

One could say that this represents an old-fashioned procedure in the face of modern commercial computer codes, but it does not for two reasons. First, technological progress is not possible without understanding the underlying problems, and the process described above helps significantly in increasing our understanding. Second also for computer codes, the users have to establish a mechanical model, and it is advisable that this model is carefully established on the basis of the thoughts discussed above. The quality of the results depends on such considerations. Commercial codes usually do not use a minimal formulation but a structure that leads to fast and efficient numerical algorithms [5, 63]. Interpretation of the results, however, depends on a thorough understanding of the mechanical model and of the real-world machine.

What we have mentioned in no way depends on the choice of the mechanical laws that were applied, NEWTON-EULER, LAGRANGE, or HAMILTON. However, the choice of the mechanical fundament for the derivation of the equations of motion considerably influences the expenditure in establishing these equations. This has to be considered very carefully; we discuss it in the following.

For the derivation of the equations of motion, we must provide a kinematic foundation, by the way one of the most frequent sources of errors and mistakes (Section 1.4). In the first step, we choose coordinate systems; this is not to be underestimated, because a good choice helps to reduce effort, a bad choice increases effort. In the second step, we determine positions, orientations, velocities, and accelerations, on the basis of these coordinate systems. In a third step, we try to find minimal coordinates, and, if necessary, we establish the constraints. Velocities and accelerations are usually needed in an absolute form and with respect to inertial and body-fixed coordinate systems. This depends on the problem under consideration. Systems with small degrees of freedom might conveniently be represented with respect to inertial coordinates, which simplifies all derivations with respect to time. The study of kinematics does not depend on the methods of kinetics, which themselves offer a broad variety of possibilities.

Momentum and moment of momentum equations according to NEWTON-EULER without application of any other mechanical principle can be preferentially used for small systems and in combination with EULER's cut principle. The result is a set of relations for each free body diagram including all reaction forces and torques. The elimination of these reaction forces without D'ALEMBERT's or JOURDAIN's principle might be cumbersome for large systems. Therefore, such a direct method makes sense only for smaller systems of clear kinematic structure.

Combining the momentum and moment of momentum equations with JOURDAIN's principle (Section 1.8) results in a very efficient method, which emerged after a very long period of discussion within the multibody systems community. The constraint forces can be eliminated by JOURDAINs principle (1.77), and the equations of motion finally include only the applied forces. The JACOBIANs of translation and rotation project the motion into the free directions as defined by the constraints. After solving these equations, we additionally can go back to certain free body diagrams for an evaluation of reaction forces (inverse kinetics). More flexibility with respect to constraint forces or contact forces, for example, is offered by LAGRANGE's equations of the first kind. In summary, we state that NEWTON-EULER together with D'ALEMBERT or JOURDAIN on the one hand, and LAGRANGE on the other hand, are presently the best procedures for treating large dynamic systems [51].

The application of LAGRANGE's equations of the second kind or of HAMILTON's principle requires evaluation of the kinetic and potential energies expressed by generalized coordinates (Section 1.9). Knowing the energies, the equations of motion follow by corresponding differentiations. This process is the most important argument against an automatized application of the analytical methods as given by LAGRANGE and HAMILTON, because computing time for differentiation is significantly greater than for the determination of the JACOBIANs for the NEWTON-EULER method. Nevertheless, analytical methods are very convenient for mechanical systems with a small number of DOF, especially in those cases where a treatment "by hand" is achievable.

The following table provides a survey of the methods presented in this chapter.

- momentum- (NEWTON) and moment of momentum (EULER) (Chapter 1.5):

$$\frac{\mathrm{d}\mathbf{p}}{\mathrm{d}t} = \mathbf{F}, \qquad \frac{\mathrm{d}\mathbf{L}_O}{\mathrm{d}t} = \mathbf{M}_O.$$

- principle of d'ALEMBERT (Section 1.7):

$$\int_K (\ddot{\mathbf{r}}\mathrm{d}m - \mathrm{d}\mathbf{F}^e)^T \, \delta\mathbf{r} = 0.$$

- principle of JOURDAIN (Section 1.7):

$$\int_K (\ddot{\mathbf{r}}\mathrm{d}m - \mathrm{d}\mathbf{F}^e)^T \, \delta\dot{\mathbf{r}} = 0.$$

- NEWTON-EULER equations (Section 1.8):

$$\left(\frac{\partial \mathbf{v}_{O'}}{\partial \dot{\mathbf{q}}}\right)^T \left[m\dot{\mathbf{v}}_{O'} + m\left(\dot{\tilde{\omega}} + \tilde{\omega}\tilde{\omega}\right)\mathbf{r}_{O'S} - \mathbf{F}^e\right]$$
$$+ \left(\frac{\partial \omega}{\partial \dot{\mathbf{q}}}\right)^T \left[m\tilde{\mathbf{r}}_{O'S}\dot{\mathbf{v}}_{O'} + \Theta_{O'}\dot{\omega} + \tilde{\omega}\Theta_{O'}\omega - \mathbf{M}^e_{O'}\right] = \mathbf{0}.$$

- LAGRANGE's equations of the first kind (Section 1.9):

$$\begin{pmatrix} \mathbf{M} & \mathbf{W} \\ \mathbf{W}^T & \mathbf{0} \end{pmatrix} \begin{pmatrix} \ddot{\mathbf{z}} \\ \lambda \end{pmatrix} + \begin{pmatrix} \mathbf{f}^g - \mathbf{f}^e \\ \bar{\mathbf{w}} \end{pmatrix} = \begin{pmatrix} \mathbf{0} \\ \mathbf{0} \end{pmatrix}.$$

- LAGRANGE's equations of the second kind (Section 1.9):

$$\left[\frac{\mathrm{d}}{\mathrm{d}t}\left(\frac{\partial T}{\partial \dot{\mathbf{q}}}\right) - \frac{\partial T}{\partial \mathbf{q}} + \frac{\partial V}{\partial \mathbf{q}}\right]^T = \mathbf{Q}_{NK}.$$

- HAMILTON's canonical equations (Section 1.10):

$$\dot{\mathbf{q}}^T = \frac{\partial H}{\partial \mathbf{p}}, \qquad \dot{\mathbf{p}}^T = -\frac{\partial H}{\partial \mathbf{q}}.$$

Example 1.14 (Robot with three revolute joints). We evaluate the equations of motion of the robot of Fig. 1.24 applying two methods, first LAGRANGE's equations of the second kind and second the NEWTON-EULER equations in minimal coordinates. According to the three revolute joints, the robot has three degrees of freedom, one rotation around a vertical axis and two rotations around horizontal axes. The two links have the lengths l_1, l_2 and the distances s_1, s_2 from their centers of mass. The joints are loaded by torque motors with the torques M_1, M_2, and M_3. The relevant acceleration of gravity is ${}_I\mathbf{g} = \begin{pmatrix} 0 & 0 & -g \end{pmatrix}^T$.

For minimal coordinates, we choose the angles $\mathbf{q} := \begin{pmatrix} q_1 & q_2 & q_3 \end{pmatrix}^T$. The body-fixed coordinates K_1 and K_2 are located at the centers of mass S_1, S_2 of the links. The necessary coordinate transformations are thus given by

$$A_{IK_1} = \begin{pmatrix} \cos q_1 \cos q_2 & -\sin q_1 & \cos q_1 \sin q_2 \\ \sin q_1 \cos q_2 & \cos q_1 & \sin q_1 \sin q_2 \\ -\sin q_2 & 0 & \cos q_2 \end{pmatrix}$$
$$= \begin{pmatrix} \cos q_1 & -\sin q_1 & 0 \\ \sin q_1 & \cos q_1 & 0 \\ 0 & 0 & 1 \end{pmatrix} \begin{pmatrix} \cos q_2 & 0 & \sin q_2 \\ 0 & 1 & 0 \\ -\sin q_2 & 0 & \cos q_2 \end{pmatrix},$$

$$A_{K_1K_2} = \begin{pmatrix} \cos q_3 & 0 & \sin q_3 \\ 0 & 1 & 0 \\ -\sin q_3 & 0 & \cos q_3 \end{pmatrix}.$$

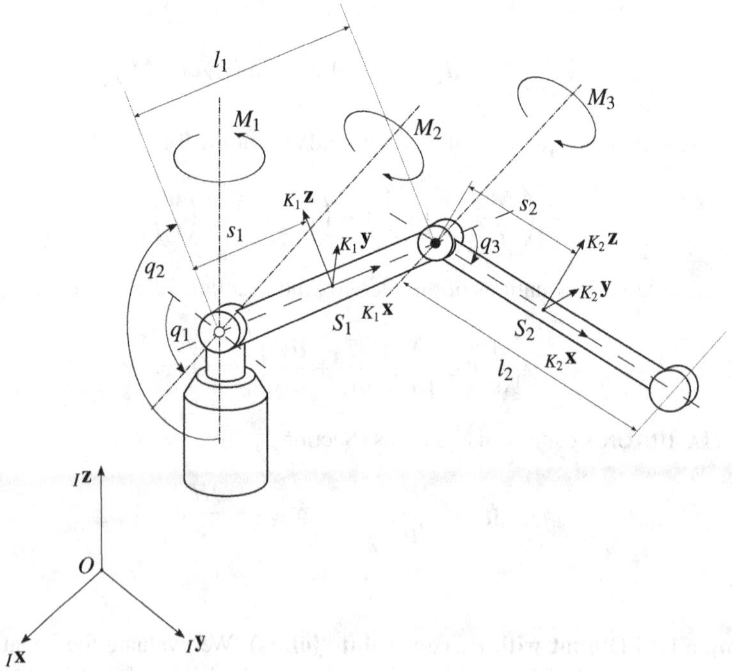

Fig. 1.24. Robot with three revolute joints.

The absolute velocities and accelerations of the two links are (in body-coordinates)

$$
{}_{K_1}\omega_1 = \begin{pmatrix} -\dot{q}_1 \sin q_2 \\ \dot{q}_2 \\ \dot{q}_1 \cos q_2 \end{pmatrix} = \begin{pmatrix} -\dot{q}_1 \sin q_2 \\ 0 \\ \dot{q}_1 \cos q_2 \end{pmatrix} + \begin{pmatrix} 0 \\ \dot{q}_2 \\ 0 \end{pmatrix},
$$

$$
{}_{K_2}\omega_2 = \begin{pmatrix} -\dot{q}_1 \sin(q_3+q_2) \\ \dot{q}_3+\dot{q}_2 \\ \dot{q}_1 \cos(q_3+q_2) \end{pmatrix} = A^T_{K_1K_2}\,{}_{K_1}\omega_1 + \begin{pmatrix} 0 \\ \dot{q}_3 \\ 0 \end{pmatrix},
$$

$$
{}_{K_1}\mathbf{v}_{S_1} = s_1 \begin{pmatrix} 0 \\ \dot{q}_1 \cos q_2 \\ -\dot{q}_2 \end{pmatrix} = {}_{K_1}\omega_1 \times \begin{pmatrix} s_1 \\ 0 \\ 0 \end{pmatrix},
$$

$$
\begin{aligned}
{}_{K_2}\mathbf{v}_{S_2} &= \begin{pmatrix} \dot{q}_2 l_1 \sin q_3 \\ \dot{q}_1 \left(l_1 \cos q_2 + s_2 \cos(q_3+q_2)\right) \\ -\dot{q}_2 l \cos q_3 - (\dot{q}_2+\dot{q}_3)\, s_2 \end{pmatrix} \\
&= {}_{K_2}\omega_2 \times \left[\begin{pmatrix} l_1 \cos q_3 \\ 0 \\ l_1 \sin q_3 \end{pmatrix} + \begin{pmatrix} s_2 \\ 0 \\ 0 \end{pmatrix} \right] + \frac{\mathrm{d}}{\mathrm{d}t} \begin{pmatrix} l_1 \cos q_3 \\ 0 \\ l_1 \sin q_3 \end{pmatrix}.
\end{aligned}
$$

With respect to the moments of inertia, we choose principal axes resulting in

$$
{}_{K_i}\Theta_{S_i,i} = \begin{pmatrix} A_i & 0 & 0 \\ 0 & B_i & 0 \\ 0 & 0 & C_i \end{pmatrix}.
$$

We have to take into account nonconservative forces and torques for both methods, for LAGRANGE's equations of the second kind and for NEWTON-EULER equations. With the JACOBIANs

$$
\left(\frac{\partial\, {}_{K_1}\omega_1}{\partial \dot{\mathbf{q}}}\right)^T = \begin{pmatrix} -\sin q_2 & 0 & \cos q_2 \\ 0 & -1 & 0 \\ 0 & 0 & 0 \end{pmatrix},
$$

$$
\left(\frac{\partial\, {}_{K_2}\omega_2}{\partial \dot{\mathbf{q}}}\right)^T = \begin{pmatrix} -\sin(q_3+q_2) & 0 & \cos(q_3+q_2) \\ 0 & -1 & 0 \\ 0 & 1 & 0 \end{pmatrix}
$$

and the torques

$$
{}_{K_1}\mathbf{M}_1 = \begin{pmatrix} -M_1 \sin q_2 \\ M_3+M_2 \\ M_1 \cos q_2 \end{pmatrix}, \quad {}_{K_2}\mathbf{M}_2 = \begin{pmatrix} 0 \\ M_3 \\ 0 \end{pmatrix},
$$

we obtain

$$\mathbf{Q}_{NK} = \left(\frac{\partial\, {}_{K_1}\omega_1}{\partial \dot{\mathbf{q}}}\right)^T {}_{K_1}\mathbf{M}_1 + \left(\frac{\partial\, {}_{K_2}\omega_2}{\partial \dot{\mathbf{q}}}\right)^T {}_{K_2}\mathbf{M}_2 = \begin{pmatrix} M_1 \\ M_2 \\ M_3 \end{pmatrix}.$$

1. **LAGRANGE*'s equations of the second kind***
   To derive the equations of motion, we need the energies. The kinetic energy T results from

$$\begin{aligned} T &= \sum_{i=1}^{2} T_i = \frac{1}{2}\sum_{i=1}^{2}\left(m_i\, {}_{K_i}\mathbf{v}_{S_i}^T\, {}_{K_i}\mathbf{v}_{S_i} + {}_{K_i}\omega_i^T\, {}_{K_i}\Theta_{S_i,i}\, {}_{K_i}\omega_i^T\right) \\ &= \frac{1}{2}m_1 s_1^2\left(\dot{q}_1^2\cos^2 q_2 + \dot{q}_2^2\right) + \frac{1}{2}\left(A_1\dot{q}_1^2\sin^2 q_2 + B_1\dot{q}_2^2 + C_1\dot{q}_1^2\cos^2 q_2\right) \\ &\quad + \frac{1}{2}m_2\left\{\dot{q}_1^2\left[l_1\cos q_2 + s_2\cos\left(q_3+q_2\right)\right]^2 + \dot{q}_2^2\left[l_1^2 + 2l_1 s_2\cos q_3\right]\right. \\ &\quad \left. +2\dot{q}_2\dot{q}_3 l_1 s_2\cos q_3 + \left(\dot{q}_3+\dot{q}_2\right)^2 s_2^2\right\} \\ &\quad + \frac{1}{2}\left\{A_2\dot{q}_1^2\sin^2\left(q_3+q_2\right) + B_2\left(\dot{q}_3+\dot{q}_2\right)^2 + C_2\dot{q}_1^2\cos^2\left(q_3+q_2\right)\right\}, \end{aligned}$$

and the potential energy is

$$V = -\sum_{i=1}^{2} m_i\, {}_I\mathbf{g}^T\, {}_I\mathbf{r}_{S_i} = -m_1 g s_1\sin q_2 - m_2 g\left[l_1\sin q_2 + s_2\sin\left(q_3+q_2\right)\right].$$

Then, the equations of motion follow from

$$\frac{\mathrm{d}}{\mathrm{d}t}\left(\frac{\partial T}{\partial \dot{\mathbf{q}}}\right) - \frac{\partial T}{\partial \mathbf{q}} + \frac{\partial V}{\partial \mathbf{q}} = \mathbf{Q}_{NK}^T$$

with

$$\begin{aligned} \frac{\partial T}{\partial \dot{q}_1} &= m_1 s_1^2\dot{q}_1\cos^2 q_2 + A_1\dot{q}_1\sin^2 q_2 + C_1\dot{q}_1\cos^2 q_2 \\ &\quad - m_2\dot{q}_2\left[l_1\cos q_2 + s_2\cos\left(q_3+q_2\right)\right]^2 \\ &\quad + A_2\dot{q}_1\sin^2\left(q_3+q_2\right) + C_2\dot{q}_1\cos^2\left(q_3+q_2\right), \\ \frac{\partial T}{\partial \dot{q}_2} &= m_1 s_1^2\dot{q}_2 - B_1\dot{q}_2 - m_2\left(l_1^2 + 2l_1 s_2\cos q_3\right)\dot{q}_2 \\ &\quad - m_2 l_1 s_2\cos q_3\dot{q}_3 - \left(\dot{q}_3+\dot{q}_2\right)s_2^2 m_2 - B_2\left(\dot{q}_3+\dot{q}_2\right), \\ \frac{\partial T}{\partial \dot{q}_3} &= m_2 l_1 s_2\cos q_3\dot{q}_2 + m_2 s_2^2\left(\dot{q}_3+\dot{q}_2\right) + B_2\left(\dot{q}_3+\dot{q}_2\right), \end{aligned}$$

$$\begin{aligned}\frac{\mathrm{d}}{\mathrm{d}t}\left(\frac{\partial T}{\partial \dot{q}_1}\right) =& \left[\left(m_1 s_1^2 + C_1\right)\cos^2 q_2 + A_1 \sin^2 q_2 + A_2 \sin^2\left(q_3 + q_2\right)\right.\\ &\left. + C_2 \cos^2\left(q_3 + q_2\right) + m_2\left(l_1 \cos q_2 + s_2 \cos\left(q_3 + q_2\right)\right)^2\right]\ddot{q}_1\\ &- \left\{-\left(A_1 - C_1 - m_1 s_1^2 - m_2 l_1^2\right)\sin 2q_2 + 2m_2 l_1 s_2 \sin\left(q_3 + 2q_2\right)\right.\\ &\left. + \left(C_2 - A_2 + m_2 s_2^2\right)\sin 2\left(q_3 + q_2\right)\right\}\dot{q}_1\dot{q}_2\\ &\left\{\left(A_2 - C_2 - m_2 s_2^2\right)\sin 2\left(q_3 + q_2\right)\right.\\ &\left. - 2m_2 l_1 s_2 \cos q_2 \sin\left(q_3 + q_2\right)\right\}\dot{q}_1\dot{q}_3,\end{aligned}$$

$$\begin{aligned}\frac{\mathrm{d}}{\mathrm{d}t}\left(\frac{\partial T}{\partial \dot{q}_2}\right) =& -\left[m_1 s_1^2 + B_1 + m_2\left(l_1^2 + 2l_1 s_2 \cos q_3\right) + m_2 s_2^2 + B_2\right]\ddot{q}_2\\ &- \left[m_2 l_1 s_2 \cos q_3 + m_2 s_2^2 + B_2\right]\ddot{q}_3 + 2m_2 l_1 s_2 \sin q_3 \dot{q}_2\dot{q}_3\\ &+ m_2 l_1 s_2 \sin q_3 \dot{q}_3^2,\end{aligned}$$

$$\begin{aligned}\frac{\mathrm{d}}{\mathrm{d}t}\left(\frac{\partial T}{\partial \dot{q}_3}\right) =& \left[m_2 l_1 s_2 \cos q_3 + m_2 s_2^2 + B_2\right]\ddot{q}_2 + \left[B_2 + m_2 s_2^2\right]\ddot{q}_3\\ &- m_2 l_1 s_2 \sin q_3 \dot{q}_2\dot{q}_3,\end{aligned}$$

$$\frac{\partial T}{\partial q_1} = 0,$$

$$\begin{aligned}\frac{\partial T}{\partial q_2} =& \left[-\frac{1}{2}A_1 \sin 2q_2 + \frac{1}{2}C_1 \sin 2q_2 + \frac{1}{2}m_1 s_1^2 \sin 2q_2\right.\\ &+ \frac{1}{2}\left(C_2 - A_2\right)\sin 2\left(q_3 + q_2\right) + \frac{1}{2}m_2 l_1^2 \sin 2q_2\\ &\left. + m_2 l_1 s_2 \sin\left(q_3 + 2q_2\right) + \frac{1}{2}m_2 s_2^2 \sin 2\left(q_3 + q_2\right)\right]\dot{q}_1^2,\end{aligned}$$

$$\begin{aligned}\frac{\partial T}{\partial q_3} =& \left[\frac{1}{2}\left(A_2 - C_2 - m_2 s_2^2\right)\sin 2\left(q_3 + q_2\right)\right.\\ &\left. - m_2 l_1 s_2 \cos q_2 \sin\left(q_3 + q_2\right)\right]\dot{q}_1^2\\ &- m_2 l_1 s_2 \sin q_3 \dot{q}_2^2 + m_2 l_1 s_2 \sin q_3 \dot{q}_2\dot{q}_3\end{aligned}$$

and

$$\frac{\partial V}{\partial q_1} = 0,$$

$$\frac{\partial V}{\partial q_2} = m_1 g s_1 \cos q_2 + m_2 g\left(l_1 \cos q_2 + s_2 \cos\left(q_3 + q_2\right)\right),$$

$$\frac{\partial V}{\partial q_3} = -m_2 g s_2 \cos\left(q_3 + q_2\right).$$

With

$$M_{11} = \left(m_1 s_1^2 + C_1\right)\cos^2 q_2 + A_1 \sin^2 q_2 + A_2 \sin^2(q_3+q_2) \\ + C_2 \cos^2(q_3+q_2) + m_2\left[l_1 \cos q_2 + s_2 \cos(q_3+q_2)\right]^2,$$

$$M_{22} = m_1 s_1^2 + m_2\left(l_1^2 + 2 l_1 s_2 \cos q_3 + s_2^2\right) + B_1 + B_2,$$

$$M_{33} = m_2 s_2^2 + B_2,$$

$$M_{23} = M_{32} = -\left[m_2\left(l_1 s_2 \cos q_3 + s_2^2\right) + B_2\right],$$

$$h_1 = -\left[-\left(A_1 - C_1 - m_1 s_1^2 - m_2 l_1^2\right)\sin 2q_2 + 2 m_2 l_1 s_2 \sin(q_3 + 2q_2) \\ + \left(m_2 s_2^2 + C_2 - A_2\right)\sin 2(q_3+q_2)\right]\dot{q}_1\dot{q}_2 + \left[\left(A_2 - C_2 - m_2 s_2^2\right)\sin 2(q_3+q_2) \\ -2 m_2 l_1 s_2 \cos q_2 \sin(q_3+q_2)\right]\dot{q}_1\dot{q}_3,$$

$$h_2 = \left[-\left(m_1 s_1^2 - A_1 + C_1 + m_2 l_1^2\right)\frac{1}{2}\sin 2q_2 + \frac{1}{2}\left(A_2 - C_2 - m_2 s_2^2\right)\sin 2(q_3+q_2) \\ - m_2 l_1 s_2 \sin(q_3+q_2)\right]\dot{q}_1^2 + m_2 l_1 s_2 \sin q_3 \dot{q}_3^2 + 2 m_2 l_1 s_2 \sin q_3 \dot{q}_2 \dot{q}_3 \\ + m_1 g s_1 \cos q_2 + m_2 g\left(l_1 \cos q_2 + s_2 \cos(q_3+q_2)\right),$$

$$h_3 = \left[m_2\left(l_1 s_2 \cos q_2 + s_2^2 \cos(q_3+q_2)\right)\sin(q_3+q_2) \\ + \frac{1}{2}(C_2 - A_2)\sin 2(q_3+q_2)\right]\dot{q}_1^2 + m_2 l_1 s_2 \sin q_3 \dot{q}_2^2 - m_2 g s_2 \cos(q_3+q_2),$$

we come to a more compact form

$$\begin{pmatrix} M_{11} & 0 & 0 \\ 0 & M_{22} & M_{23} \\ 0 & M_{23} & M_{33} \end{pmatrix}\ddot{\mathbf{q}} + \begin{pmatrix} h_1 \\ h_2 \\ h_3 \end{pmatrix} = \begin{pmatrix} M_1 \\ M_2 \\ M_3 \end{pmatrix}.$$

2. **NEWTON-EULER** ***equations***
Using the cut principle, we apply the momentum and moment of momentum equations to every body of the robot and get:

$$\sum_{i=1}^{2}\left\{\left(\frac{\partial\, {}_{K_i}\mathbf{v}_{S_i}}{\partial \dot{\mathbf{q}}}\right)^T \left(m_i\, {}_{K_i}\dot{\mathbf{v}}_{S_i} - {}_{K_i}\mathbf{F}_i^e\right) \right. \\ \left. + \left(\frac{\partial\, {}_{K_i}\omega_i}{\partial \dot{\mathbf{q}}}\right)^T \left({}_{K_i}\Theta_{S_i,i}\, {}_{K_i}\dot{\omega}_i + {}_{K_i}\tilde{\omega}_i\, {}_{K_i}\Theta_{S_i,i}\, {}_{K_i}\omega - {}_{K_i}\mathbf{M}_i\right)\right\} = \mathbf{0}.$$

The rotational and translational accelerations are

$$
{}_{K_1}\dot{\mathbf{v}}_{S_1} = s_1 \begin{pmatrix} l - \dot{q}_1^2 \cos^2 q_2 - \dot{q}_2^2 \\ \ddot{q}_1 \cos q_2 - 2\dot{q}_1 \dot{q}_2 \sin q_2 \\ -\ddot{q}_2 - \dot{q}_1^2 \sin q_2 \cos q_2 \end{pmatrix},
$$

$$
\begin{aligned}
{}_{K_2}\dot{\mathbf{v}}_{S_2} = & \begin{pmatrix} -\ddot{q}_2 l_1 \sin q_3 - \dot{q}_1^2 \left[l_1 \cos q_2 + s_2 \cos(q_3+q_2)\right] \cos(q_3+q_2) \\ \ddot{q}_1 \left[l_1 \cos q_2 + s_2 \cos(q_3+q_2)\right] - 2\dot{q}_1 \dot{q}_2 l_1 \sin q_2 \\ -\ddot{q}_2 (l_1 \cos q_3 + s_2) - \ddot{q}_3 s_2 - \dot{q}_1^2 \left[l_1 \cos q_2 + s_2 \cos(q_3+q_2)\right] \sin(q_3+q_2) \end{pmatrix} \\
& + \begin{pmatrix} -\dot{q}_2^2 (l_1 \cos q_3 + s_2) - \dot{q}_3^2 s_2 - 2\dot{q}_2 \dot{q}_3 s_2 \\ -2\dot{q}_1 \dot{q}_2 s_2 \sin(q_3+q_2) - 2\dot{q}_1 \dot{q}_3 s_2 \sin(q_3+q_2) \\ -\dot{q}_2^2 l_1 \sin q_3 \end{pmatrix},
\end{aligned}
$$

$$
{}_{K_1}\dot{\omega}_1 = \begin{pmatrix} -\ddot{q}_1 \sin q_2 - \dot{q}_1 \dot{q}_2 \cos q_2 \\ \ddot{q}_2 \\ \ddot{q}_1 \cos q_2 - \dot{q}_1 \dot{q}_2 \sin q_2 \end{pmatrix},
$$

$$
{}_{K_2}\dot{\omega}_2 = \begin{pmatrix} -\ddot{q}_1 \sin(q_3+q_2) - \dot{q}_1 (\dot{q}_3+\dot{q}_2) \cos(q_3+q_2) \\ \ddot{q}_3 + \ddot{q}_2 \\ \ddot{q}_1 \cos(q_3+q_2) - \dot{q}_1 (\dot{q}_3+\dot{q}_2) \sin(q_3+q_2) \end{pmatrix}.
$$

From this, we come to the JACOBIANs of rotation and translation

$$
\left(\frac{\partial\, {}_{K_1}\mathbf{v}_{S_1}}{\partial \dot{\mathbf{q}}}\right)^T = \begin{pmatrix} 0 & s_1 \cos q_2 & 0 \\ 0 & 0 & s_1 \\ 0 & 0 & 0 \end{pmatrix},
$$

$$
\left(\frac{\partial\, {}_{K_2}\mathbf{v}_{S_2}}{\partial \dot{\mathbf{q}}}\right)^T = \begin{pmatrix} 0 & l_1 \cos q_2 + s_2 (q_3+q_2) & 0 \\ -l_1 \sin q_3 & 0 & l_1 \cos q_3 + s_2 \\ 0 & 0 & -s_2 \end{pmatrix}.
$$

The applied forces write

$$
{}_{K_1}\mathbf{F}_1^e = -m_1 g \begin{pmatrix} -\sin q_2 \\ 0 \\ \cos q_2 \end{pmatrix},
$$

$$
{}_{K_2}\mathbf{F}_2^e = -m_2 g \begin{pmatrix} -\sin(q_3+q_2) \\ 0 \\ \cos(q_3+q_2) \end{pmatrix}.
$$

Combining all this by the use of the NEWTON-EULER equations, we finally come out with the same results as compared with LAGRANGE's equations of the second kind.

Chapter 2
Linear Discrete Models

2.1 Linearization

In the presentation of the methods in Chapter 1, we have already assumed concrete conceptual models, that is rigid bodies with homogeneous, constant mass that are arbitrarily connected by constraints. The motion of such multibody systems with f degrees of freedom is described by ordinary and usually nonlinear second-order differential equations with (minimal) form

$$\mathbf{M}(\mathbf{q},t)\ddot{\mathbf{q}} = \mathbf{h}(\mathbf{q},\dot{\mathbf{q}},\mathbf{Q},t) \ . \tag{2.1}$$

Here $\mathbf{q}(t)$ is the vector of minimal coordinates and $\mathbf{M}$ an always symmetric positive-definite mass matrix. The vector $\mathbf{h}$ contains gyroscopic and dissipative forces, as well as all applied forces and moments. The vector $\mathbf{Q}$ contains applied forces and moments, which we would like to parametrize. In many practically relevant cases, the quantities $\mathbf{M}$ and $\mathbf{h}$ do not depend explicitly on time.

According to the previous model considerations, we assume, first, that a *reference motion* or a stationary state for the system's motion can be found, and, second, that small oscillations occur about this reference. Then, the vector of minimal coordinates $\mathbf{q}(t)$ can be split into a *reference vector* $\mathbf{q}_0(t)$ and a *perturbation vector* $\eta_{\mathbf{q}}(t)$ [42]. Similarly, we split the vector $\mathbf{Q}$:

$$\mathbf{q}(t) = \mathbf{q}_0(t) + \eta_{\mathbf{q}}(t) \ , \tag{2.2}$$

$$\mathbf{Q}(t) = \mathbf{Q}_0(t) + \eta_{\mathbf{Q}}(t) \ . \tag{2.3}$$

We insert these expressions into the original equation of motion (2.1) and expand the latter as a TAYLOR series at $\mathbf{q}_0(t)$, $\dot{\mathbf{q}}_0(t)$, $\ddot{\mathbf{q}}_0(t)$ and $\mathbf{Q}_0(t)$ with

F. Pfeiffer and T. Schindler, *Introduction to Dynamics*,
DOI: 10.1007/978-3-662-46721-3_2

$$\mathbf{M}\left((\mathbf{q}_0+\eta_{\mathbf{q}}),t\right)=\mathbf{M}(\mathbf{q}_0,t)+\sum_{i=1}^{f}\left(\frac{\partial \mathbf{M}}{\partial q_i}\right)_0 \eta_{\mathbf{q}_i}+\text{hot}\,, \tag{2.4}$$

$$\begin{aligned}\mathbf{h}\left((\mathbf{q}_0+\eta_{\mathbf{q}}),(\dot{\mathbf{q}}_0+\dot{\eta}_{\mathbf{q}}),(\mathbf{Q}_0+\eta_{\mathbf{Q}}),t\right)&=\mathbf{h}(\mathbf{q}_0,\dot{\mathbf{q}}_0,\mathbf{Q}_0,t)\\&+\left(\frac{\partial \mathbf{h}}{\partial \mathbf{q}}\right)_0\eta_{\mathbf{q}}+\left(\frac{\partial \mathbf{h}}{\partial \dot{\mathbf{q}}}\right)_0\dot{\eta}_{\mathbf{q}}+\left(\frac{\partial \mathbf{h}}{\partial \mathbf{Q}}\right)_0\eta_{\mathbf{Q}}+\text{hot}\,.\end{aligned} \tag{2.5}$$

The abbreviation *hot* stands for higher order terms. According to the rules of vector and tensor analysis [49]

- the derivative of a scalar with respect to a vector results in a row vector,
- the derivative of a vector with respect to a vector results in a second-order tensor (JACOBIAN matrix),
- the derivative of a second-order tensor (mass matrix) with respect to a vector results in a third-order tensor.

Assuming that also $\dot{\eta}_{\mathbf{q}}$ and $\ddot{\eta}_{\mathbf{q}}$ are small, we obtain two equations of motion, one for the reference motion and one for the linearized *perturbation*:

- reference motion

$$\mathbf{M}(\mathbf{q}_0,t)\,\ddot{\mathbf{q}}_0=\mathbf{h}(\mathbf{q}_0,\dot{\mathbf{q}}_0,\mathbf{Q}_0,t)\,, \tag{2.6}$$

- perturbation (linearized)

$$\mathbf{M}(\mathbf{q}_0,t)\,\ddot{\eta}_{\mathbf{q}}+\mathbf{P}(\mathbf{q}_0,\dot{\mathbf{q}}_0,\mathbf{Q}_0,t)\,\dot{\eta}_{\mathbf{q}}+\mathbf{R}(\mathbf{q}_0,\dot{\mathbf{q}}_0,\ddot{\mathbf{q}}_0,\mathbf{Q}_0,t)\,\eta_{\mathbf{q}}=\mathbf{f} \tag{2.7}$$

with the right-hand side

$$\mathbf{f}=\left(\frac{\partial \mathbf{h}}{\partial \mathbf{Q}}\right)_0\eta_{\mathbf{Q}}\,, \tag{2.8}$$

the matrix for velocity-depending forces

$$\mathbf{P}(\mathbf{q}_0,\dot{\mathbf{q}}_0,\mathbf{Q}_0,t)=-\left(\frac{\partial \mathbf{h}}{\partial \dot{\mathbf{q}}}\right)_0\,, \tag{2.9}$$

as well as the matrix for position-depending forces

$$\mathbf{R}(\mathbf{q}_0,\dot{\mathbf{q}}_0,\ddot{\mathbf{q}}_0,\mathbf{Q}_0,t)=\left(\frac{\partial \mathbf{M}}{\partial \mathbf{q}}\right)_0\ddot{\mathbf{q}}_0-\left(\frac{\partial \mathbf{h}}{\partial \mathbf{q}}\right)_0\,. \tag{2.10}$$

2.2 Classification of Linear Systems

The above splitting into reference and perturbation usually results in a nonlinear system of differential equations for the reference motion (2.6), and a linear matrix-vector system for the perturbation (2.7). In most practical cases, the reference motion is known or it may be determined from the nonlinear equations. Then, the perturbation about such a reference motion will be of interest. We consider it in the following. It is worthwhile, in a first step, to take a closer look at the structure of the homogeneous linearized equations of motion (2.7) ($\mathbf{f} = \mathbf{0}$). The matrices $\mathbf{P}$ and $\mathbf{R}$ can always be decomposed into a symmetric and a skew-symmetric part:

$$\mathbf{D} = \frac{1}{2}\left(\mathbf{P} + \mathbf{P}^T\right) = \mathbf{D}^T \,, \tag{2.11}$$

$$\mathbf{G} = \frac{1}{2}\left(\mathbf{P} - \mathbf{P}^T\right) = -\mathbf{G}^T \,, \tag{2.12}$$

$$\mathbf{K} = \frac{1}{2}\left(\mathbf{R} + \mathbf{R}^T\right) = \mathbf{K}^T \,, \tag{2.13}$$

$$\mathbf{N} = \frac{1}{2}\left(\mathbf{R} - \mathbf{R}^T\right) = -\mathbf{N}^T \,. \tag{2.14}$$

Thus, we obtain from (2.7):

$$\mathbf{M}\ddot{\eta}_{\mathbf{q}} + (\mathbf{D} + \mathbf{G})\,\dot{\eta}_{\mathbf{q}} + (\mathbf{K} + \mathbf{N})\,\eta_{\mathbf{q}} = \mathbf{f} \,. \tag{2.15}$$

The single matrices have the following properties:

M mass matrix (symmetric)
The term $\mathbf{M}\ddot{\eta}_{\mathbf{q}}$ represents the inertia forces. It follows from the kinetic energy of the perturbation $T = \frac{1}{2}\dot{\eta}_{\mathbf{q}}^T\mathbf{M}\dot{\eta}_{\mathbf{q}}$.

D damping matrix (symmetric)
The term $\mathbf{D}\dot{\eta}_{\mathbf{q}}$ represents damping forces that are proportional to the velocity. It follows from RAYLEIGH's damping power $\frac{1}{2}\dot{\eta}_{\mathbf{q}}^T\mathbf{D}\dot{\eta}_{\mathbf{q}}$.

G gyroscopic matrix (skew-symmetric)
The expression $\mathbf{G}\dot{\eta}_{\mathbf{q}}$ contains gyroscopic forces. The power $\dot{\eta}_{\mathbf{q}}^T\mathbf{G}\dot{\eta}_{\mathbf{q}} = 0$ vanishes due to the skew-symmetry of $\mathbf{G}$. Gyroscopic forces do not change the energy balance of the system [46].

K stiffness matrix (symmetric)
The expression $\mathbf{K}\eta_{\mathbf{q}}$ contains the conservative perturbation forces (position-depending forces). It can be derived from the potential of the perturbation $V = \frac{1}{2}\eta_{\mathbf{q}}^T\mathbf{K}\eta_{\mathbf{q}}$.

N circulatory matrix (skew-symmetric)
The circulatory matrix defines nonconservative forces $\mathbf{N}\eta_{\mathbf{q}}$, which occur for example in turbines and bearings [72]. On the one hand, the power $\dot{\eta}_{\mathbf{q}}\mathbf{N}\eta_{\mathbf{q}}$ does not vanish, on the other hand $\dot{\eta}_{\mathbf{q}}\mathbf{N}\eta_{\mathbf{q}}$ is not symmetric. The integral

$$\int_{\eta_{q1}}^{\eta_{q2}} \eta_{\mathbf{q}}^T \mathbf{N} \mathrm{d}\eta_{\mathbf{q}} = \int_{t_1}^{t_2} \eta_{\mathbf{q}}^T \mathbf{N} \dot{\eta}_{\mathbf{q}} \mathrm{d}t$$

depends on the path of integration and therefore on the choice of the minimal coordinates (Section 1.6).

If the damping forces ($\mathbf{D} = \mathbf{0}$) and the nonconservative position-depending forces ($\mathbf{N} = \mathbf{0}$) do not exist, we get a conservative system, for which the total energy is constant and thus the energy conservation law is valid (Section 1.6). The classification of discrete linear systems concerning these physically interpretable matrices not only has considerable advantages for the mathematical and numerical treatment, but also gives direct qualitative insight about the system's behavior. This is important and helpful in particular for stability predictions [45].

For the further treatment of systems of differential equations, it is often convenient to transform f linear second-order differential equations to $2f$ linear first-order differential equations in state space form. By means of the substitution

$$\mathbf{x} = \begin{pmatrix} \eta_{\mathbf{q}} \\ \dot{\eta}_{\mathbf{q}} \end{pmatrix} , \tag{2.16}$$

we obtain

$$\dot{\mathbf{x}} = \mathbf{A}(t)\mathbf{x} + \mathbf{b}(t) \tag{2.17}$$

with

$$\mathbf{A} = \begin{pmatrix} \mathbf{0} & \mathbf{E} \\ -\mathbf{M}^{-1}\mathbf{R} & -\mathbf{M}^{-1}\mathbf{P} \end{pmatrix} , \quad \mathbf{b} = \begin{pmatrix} \mathbf{0} \\ \mathbf{M}^{-1}\mathbf{f} \end{pmatrix} . \tag{2.18}$$

For this formulation, standard solution methods exist, which we will study in the following. The vector $\mathbf{x}$ is called the *state vector*. Accordingly, we speak of the *state space* with the coordinates $\mathbf{x}$, whereas the coordinates $\eta_{\mathbf{q}}$ represent the configuration space (Section 1.4).

Example 2.1 (Spherical pendulum). We consider the spherical pendulum in Fig. 2.1. In Example 1.12, the equations of motion, those of the cyclic coordinates included, have been derived with the minimal coordinates $\mathbf{q} = (\psi, \vartheta)^T$:

$$\begin{aligned} mR^2 \sin^2 \vartheta \ddot{\psi} + 2mR^2 \sin \vartheta \cos \vartheta \dot{\vartheta} \dot{\psi} &= 0 , \\ mR^2 \ddot{\vartheta} - mR^2 \sin \vartheta \cos \vartheta \dot{\psi}^2 + mgR \sin \vartheta &= 0 , \\ \sin^2 \vartheta \dot{\psi} &= C . \end{aligned}$$

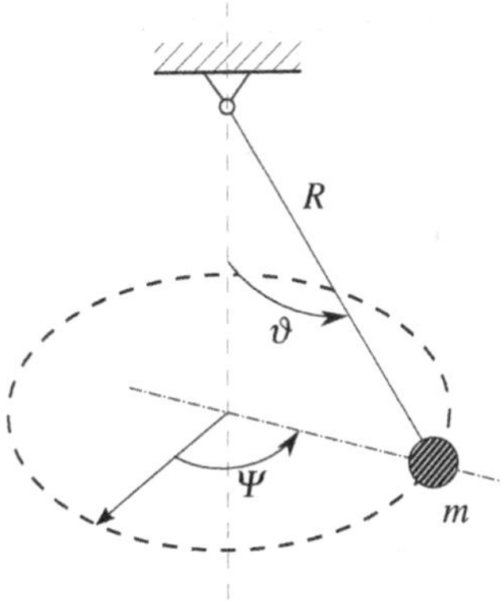

Fig. 2.1. Spherical pendulum.

- First, we consider the reduction to the planar motion ($\dot{\psi} \equiv 0$)

$$\ddot{\vartheta} + \frac{g}{R} \sin \vartheta = 0 \,.$$

The condition $\vartheta_0 = 0$ defines the *stationary state* of the system and is obtained from $\ddot{\mathbf{q}} = \dot{\mathbf{q}} = \mathbf{0}$. We define the stationary state as the reference motion and we are interested in the plane motion of the pendulum for small deflections

$$\vartheta = \underbrace{\vartheta_0}_{=0} + \eta_\vartheta \,.$$

With $\sin \vartheta = \vartheta - \frac{1}{6}\vartheta^3 + \text{hot}$, the perturbation satisfies the linear differential equation

$$\ddot{\vartheta} + \frac{g}{R} \vartheta = 0 \,.$$

- Another reference motion is the cone-shaped path ($\vartheta \equiv \vartheta_0 \neq 0$). Depending on the choice as to which of the original differential equations is used, one obtains the relations

$$\dot{\psi}^2 = \frac{C^2}{\sin^4 \vartheta_0} = \frac{g}{R \cos \vartheta_0} \,.$$

With the moment equilibrium about P (Fig. 2.2)

$$mgR \sin \vartheta_0 - m a_r R \cos \vartheta_0 = 0 \,,$$

we get

$$g \sin \vartheta_0 - \dot{\psi}^2 R \sin \vartheta_0 \cos \vartheta_0 = 0$$

and the physical interpretation of the reference motion:

$$\dot{\psi}^2 = \frac{g}{R\cos\vartheta_0}\,.$$

We are interested in a perturbation in the direction of

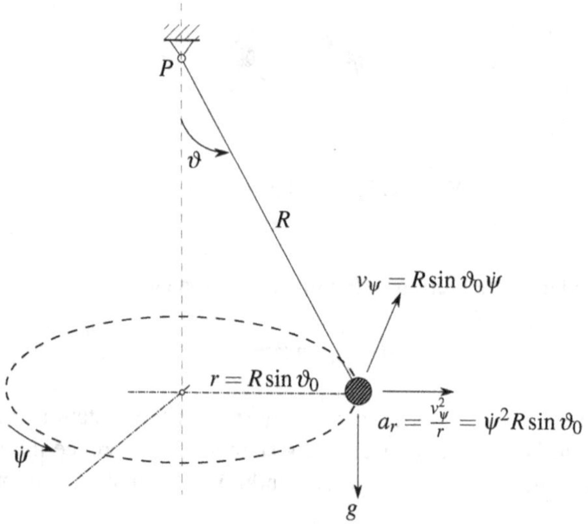

Fig. 2.2. Moment equilibrium for the spherical pendulum.

$$\vartheta = \vartheta_0 + \eta_\vartheta$$

and derive the corresponding differential equation for η_ϑ. We use $\dot{\psi} = \frac{C}{\sin^2\vartheta}$ in the second original differential equation:

$$\ddot{\vartheta} - C^2\left(\frac{\cos\vartheta}{\sin^3\vartheta}\right) + \frac{g}{R}\sin\vartheta = 0\,.$$

The differential equation is of the type

$$\ddot{\vartheta} + f(\vartheta) = 0\,.$$

With

$$\sin\vartheta \doteq \sin\vartheta_0 + \eta_\vartheta\cos\vartheta_0\,,$$
$$\cos\vartheta \doteq \cos\vartheta_0 - \eta_\vartheta\sin\vartheta_0\,,$$

we obtain

$$\begin{aligned}
0 \doteq & \underbrace{\ddot{\vartheta}_0}_{=0} + \ddot{\eta}_\vartheta + \underbrace{f(\vartheta_0)}_{=0} + \left(\frac{\partial f}{\partial \vartheta}\right)_0 \eta_\vartheta \\
&= \ddot{\eta}_\vartheta + \left\{ -\frac{C^2}{\sin^6 \vartheta}\left[-\sin^4\vartheta - 3\sin^2\vartheta\cos^2\vartheta\right] + \left(\frac{g}{R}\right)\cos\vartheta \right\}_0 \eta_\vartheta \\
&= \ddot{\eta}_\vartheta + \left(\frac{g}{R\cos\vartheta_0}\right)\left(\sin^2\vartheta_0 + 3\cos^2\vartheta_0 + \cos^2\vartheta_0\right)\eta_\vartheta \\
&= \ddot{\eta}_\vartheta + \left[\left(\frac{g}{R\cos\vartheta_0}\right)\left(1 + 3\cos^2\vartheta_0\right)\right]\eta_\vartheta \,.
\end{aligned}$$

Therefore, it is

$$\omega = \sqrt{\left(\frac{g}{R}\right)\left(\frac{1}{\cos\vartheta_0} + 3\cos\vartheta_0\right)}$$

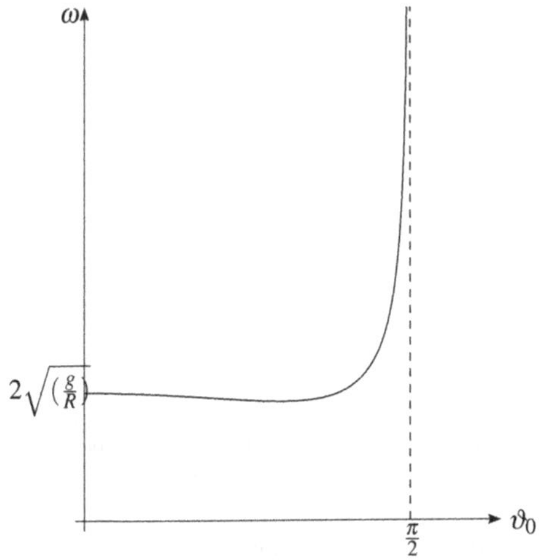

Fig. 2.3. Eigen angular frequency of the perturbation for the cone-shaped reference.

the *eigen angular frequency* of the perturbation (Fig. 2.3). We consider some special cases for ϑ_0:

1. $\vartheta_0 \ll 1 : \omega \approx 2\sqrt{\frac{g}{R}}$ and $\dot{\psi} \approx \sqrt{\frac{g}{R}}$
   Because of the perturbation, the originally circular orbit for ψ is deformed to an ellipse (Fig. 2.4a).

2. $\vartheta_0 \to \frac{\pi}{2} : \omega = \dot{\psi} \to \infty$
   The equatorial curve is somewhat inclined (Fig. 2.4b).

The case $\vartheta_0 > \frac{\pi}{2}$ is not possible, otherwise $\dot{\psi}^2 = \frac{g}{R\cos\vartheta_0}$ would become negative.

3. $0 < \vartheta_0 < \frac{\pi}{2}$
As for $\vartheta_0 \ll 1$, one obtains ellipse-shaped curves, which repeat periodically (Fig. 2.4c) with $2\dot{\psi}_0 > \omega > \dot{\psi}_0$. If ϑ_0 is not too large, one can interpret this as an ellipse with a rotating principal axis. For two complete oscillations with period T, the marching angle ψ^* satisfies

$$\psi^* = \int_0^{2T} \dot{\psi}\mathrm{d}t - 2\pi \approx \dot{\psi}_0 2T - 2\pi\,.$$

With

$$2T = \left(\frac{4\pi}{\omega}\right) = \frac{4\pi}{\dot{\psi}_0\sqrt{1+3\cos^2\vartheta_0}}\,,$$

we get

$$\psi^* \approx 2\pi\left(\frac{2}{\sqrt{1+3\cos^2\vartheta_0}} - 1\right).$$

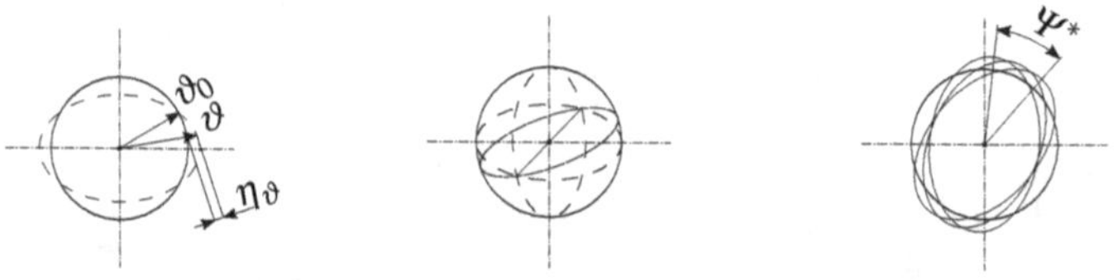

(a) $\vartheta_0 \ll 1$: curve of the perturbed cone-shaped path (seen from above).

(b) $\vartheta_0 \to \frac{\pi}{2}$: curve of the perturbed cone-shaped path.

(c) $0 < \vartheta_0 < \frac{\pi}{2}$: curve of the perturbed cone-shaped path.

Fig. 2.4. Limiting cases for the curve of the perturbed cone-shaped path.

2.3 Solution Methods

With the selection and definition of a model as well as the mathematical description of this model in the form of differential equations, we have a mechanical and mathematical model for the system. The equations of motion are the basis for all further investigations. For the evaluation and interpretation of a dynamic system, we are interested in the following aspects and problems:

- Motion (time response, frequency response): motion, motion patterns, frequency, damping, stability, amplitude, and phase response functions.

- Control: If the system is controlled, questions of observability, controllability, control quality, control stability, and control optimization have to be answered.
- Perturbation: perturbation of the system, sensitivity of parameters, deterministic and stochastic perturbation.
- Optimization: Optimization of the dynamic system as a whole (process + controller) with respect to certain performance criteria, design strategies for the optimization of parameters and structures regarding sensitivities or other criteria.

The necessary methods involve a wide range of mathematics, system and control theory. They have multidisciplinary character. In the context of this paper, only the most important methods can be discussed.

2.3.1 Linear Second-Order Systems

In this section, we consider the original second-order differential equation and restrict ourselves to the typical cases, simple eigenvalues and periodic excitation. The more general case is discussed in Section 2.3.2.

We first restrict ourselves to conservative mechanical systems without a right-hand side:

$$\mathbf{M}\ddot{\eta}_{\mathbf{q}} + \mathbf{K}\eta_{\mathbf{q}} = \mathbf{0}\,, \quad \eta_{\mathbf{q}} \in \mathrm{I\!R}^{f}\,, \quad \{\mathbf{M}, \mathbf{K}\} \in \mathrm{I\!R}^{f,f}\,. \tag{2.19}$$

The matrices $\mathbf{M}$ and $\mathbf{K}$ are assumed to be constant. For the solution vector, we choose the approach

$$\eta_{\mathbf{q}} = \bar{\eta}_{\mathbf{q}} e^{\lambda t}\,. \tag{2.20}$$

We obtain a homogeneous system of equations for the vector $\bar{\eta}_{\mathbf{q}}$:

$$\left(\lambda^2 \mathbf{M} + \mathbf{K}\right) \bar{\eta}_{\mathbf{q}} = \mathbf{0}\,. \tag{2.21}$$

The homogeneous system has a nontrivial solution if and only if the determinant vanishes

$$P(\lambda) := \det\left(\lambda^2 \mathbf{M} + \mathbf{K}\right) = 0\,, \tag{2.22}$$

which defines a *characteristic equation* for its zeros, the so-called *eigenvalues* $\{\lambda_i\}_i$, using the *characteristic polynomial* $P(\lambda)$ [65]. Analogously, the eigenvalues of a homogeneous linear second-order differential equation with damping can be determined:

$$P(\lambda) := \det\left(\lambda^2 \mathbf{M} + \lambda\,(\mathbf{D} + \mathbf{G}) + (\mathbf{K} + \mathbf{N})\right) = 0\,. \tag{2.23}$$

One always obtains as many complex conjugate eigenvalue pairs as degrees of freedom. In the case of (2.19), the eigenvalues are even purely imaginary [65, 46]:

$$\lambda_i = \pm j\omega_i\,. \tag{2.24}$$

The *eigenvectors* $\{\bar{\eta}_{\mathbf{q}_i}\}_i$ are derived from the eigenvalues up to a factor

$$\left(\lambda_i^2\mathbf{M}+\lambda_i(\mathbf{D}+\mathbf{G})+(\mathbf{K}+\mathbf{N})\right)\bar{\eta}_{\mathbf{q}_i}=\mathbf{0}\,, \tag{2.25}$$

where we assume that we can find f linearly independent complex conjugate eigenvector pairs with this procedure. Complex conjugate eigenvalues always belong to complex conjugate eigenvectors [65]:

$$\lambda_i=\delta_i+j\omega_i \quad\Rightarrow\quad \bar{\eta}_{\mathbf{q}_i}=\alpha_i+j\beta_i\,, \tag{2.26}$$

$$\lambda_{i+f}=\delta_i-j\omega_i \quad\Rightarrow\quad \bar{\eta}_{\mathbf{q}_{i+f}}=\alpha_i-j\beta_i\,. \tag{2.27}$$

The mapping λ_i to λ_{i+f} is a question of definition. In case of (2.19) with $\delta_i=0$, we obtain real eigenvectors $\bar{\eta}_{\mathbf{q}_i}=\alpha_i$ (no phase shift). For conservative vibration systems, these eigenvectors represent the *mode shapes*; in general, one obtains only *fundamental vibrations*. The solution of the homogeneous system is obtained as a linear combination of all fundamental vibrations:

$$\eta_{\mathbf{q}}(t)=\sum_{i=1}^{f}e^{\delta_i t}\left(c_i\bar{\eta}_{\mathbf{q}_i}e^{j\omega_i t}+c_{i+f}\bar{\eta}_{\mathbf{q}_{i+f}}e^{-j\omega_i t}\right)\,. \tag{2.28}$$

Without damping, all eigenvectors are real and occur twice; c_{i+f} is complex conjugate to c_i. This gives the real representation

$$\eta_{\mathbf{q}}(t)=\sum_{i=1}^{f}\bar{\eta}_{\mathbf{q}_i}\left(a_i\cos(\omega_i t)+b_i\sin(\omega_i t)\right) \tag{2.29}$$

with $a_i=2\Re(c_i)$ and $b_i=-2\Im(c_i)$.

With the help of the *modal matrix*

$$\mathbf{V}=\left(\bar{\eta}_{\mathbf{q}_1},\bar{\eta}_{\mathbf{q}_2},\ldots,\bar{\eta}_{\mathbf{q}_f}\right) \tag{2.30}$$

from (2.29), we get the matrix-vector relationship

$$\eta_{\mathbf{q}}=\mathbf{V}\left[\begin{pmatrix}\cos(\omega_1 t) & 0 & \cdots & 0\\ 0 & \ddots & \ddots & \vdots\\ \vdots & \ddots & \ddots & 0\\ 0 & \cdots & 0 & \cos(\omega_f t)\end{pmatrix}\begin{pmatrix}a_1\\ \vdots\\ a_f\end{pmatrix} + \begin{pmatrix}\sin(\omega_1 t) & 0 & \cdots & 0\\ 0 & \ddots & \ddots & \vdots\\ \vdots & \ddots & \ddots & 0\\ 0 & \cdots & 0 & \sin(\omega_f t)\end{pmatrix}\begin{pmatrix}b_1\\ \vdots\\ b_f\end{pmatrix}\right]\,. \tag{2.31}$$

We write

$$\eta_{\mathbf{q}} = \mathbf{V}\left[\cos(\Omega t)\,\mathbf{a} + \sin(\Omega t)\,\mathbf{b}\right] \tag{2.32}$$

with

$$\cos(\Omega t) = \operatorname{diag}\{\cos(\omega_i t)\}\ , \quad \sin(\Omega t) = \operatorname{diag}\{\sin(\omega_i t)\}\ , \tag{2.33}$$

$$\mathbf{a} = \left(a_1, \cdots, a_f\right)^T\ , \quad \mathbf{b} = \left(b_1, \cdots, b_f\right)^T\ . \tag{2.34}$$

With the *initial conditions* $\eta_{\mathbf{q}} = \eta_{\mathbf{q}_0}$ and $\dot{\eta}_{\mathbf{q}} = \dot{\eta}_{\mathbf{q}_0}$ at $t = 0$, the unknown vectors **a** and **b** can be defined:

$$\eta_{\mathbf{q}_0} = \mathbf{V}\mathbf{a}\ , \quad \dot{\eta}_{\mathbf{q}_0} = \mathbf{V}\Omega\mathbf{b}\ , \quad \text{with } \Omega = \operatorname{diag}\{\omega_i\}\ . \tag{2.35}$$

Because of the invertibility of **V**, it follows

$$\mathbf{a} = \mathbf{V}^{-1}\eta_{\mathbf{q}_0}\ , \tag{2.36}$$

$$\mathbf{b} = \Omega^{-1}\mathbf{V}^{-1}\dot{\eta}_{\mathbf{q}_0} \tag{2.37}$$

and therefore without damping:

$$\eta_{\mathbf{q}}(t) = \mathbf{V}\cos(\Omega t)\,\mathbf{V}^{-1}\eta_{\mathbf{q}_0} + \mathbf{V}\sin(\Omega t)\,\Omega^{-1}\mathbf{V}^{-1}\dot{\eta}_{\mathbf{q}_0}\ . \tag{2.38}$$

The initial conditions determine the components of the vectors **a** and **b** and thus the eigen angular frequencies. The modal matrix **V** acts as a distribution among the individual degrees of freedom using the eigenvectors, which are arranged column-by-column correspondingly. The practical significance of the eigenvectors $\bar{\eta}_{\mathbf{q}_1}, \ldots, \bar{\eta}_{\mathbf{q}_f}$ is often underestimated. The distribution among the individual degrees of freedom gives an idea which components oscillate and which do not and which components vibrate against each other, especially for forced oscillations. This is essential for large mechanical systems and allows conclusions for possible design improvements.

We use the solution (2.38) in the differential equation (2.19). For simplicity, let $\dot{\eta}_{\mathbf{q}_0} = \mathbf{0}$. With

$$\ddot{\eta}_{\mathbf{q}} = -\mathbf{V}\Omega^2\cos(\Omega t)\,\mathbf{V}^{-1}\eta_{\mathbf{q}_0} \tag{2.39}$$

and

$$\mathbf{M}^{-1}\mathbf{K}\eta_{\mathbf{q}} = \mathbf{M}^{-1}\mathbf{K}\mathbf{V}\cos(\Omega t)\,\mathbf{V}^{-1}\eta_{\mathbf{q}_0}\ , \tag{2.40}$$

it follows

$$\begin{aligned}\mathbf{0} &= -\mathbf{V}\Omega^2\cos(\Omega t)\,\mathbf{V}^{-1}\eta_{\mathbf{q}_0} + \mathbf{M}^{-1}\mathbf{K}\mathbf{V}\cos(\Omega t)\,\mathbf{V}^{-1}\eta_{\mathbf{q}_0} \\ &= \left[-\mathbf{V}\Omega^2 + \mathbf{M}^{-1}\mathbf{K}\mathbf{V}\right]\cos(\Omega t)\,\mathbf{V}^{-1}\eta_{\mathbf{q}_0}\end{aligned} \tag{2.41}$$

As $\eta_{\mathbf{q}_0}$ is arbitrary, it is

$$\Omega^2 = \mathbf{V}^{-1}\mathbf{M}^{-1}\mathbf{K}\mathbf{V}\ . \tag{2.42}$$

Conversely, the equations of motion

$$\ddot{\boldsymbol{\eta}}_{\mathbf{q}} + \mathbf{M}^{-1}\mathbf{K}\boldsymbol{\eta}_{\mathbf{q}} = \mathbf{0} \tag{2.43}$$

can be transformed with the *modal transformation*

$$\boldsymbol{\eta}_{\mathbf{q}} = \mathbf{V}\xi \quad \text{or} \quad \xi = \mathbf{V}^{-1}\boldsymbol{\eta}_{\mathbf{q}} \tag{2.44}$$

to

$$\begin{aligned}\mathbf{0} &= \mathbf{V}\ddot{\xi} + \mathbf{M}^{-1}\mathbf{K}\mathbf{V}\xi = \ddot{\xi} + \mathbf{V}^{-1}\mathbf{M}^{-1}\mathbf{K}\mathbf{V}\xi \\ &= \ddot{\xi} + \Omega^2\xi\,. \end{aligned} \tag{2.45}$$

The modal transformation causes a *decoupling* of f equations, such that a single scalar equation, which is not coupled with other modes of vibration, results for each oscillation frequency. We call ξ *modal coordinates* as opposed to the *natural coordinates* $\boldsymbol{\eta}_{\mathbf{q}}$ and speak of *fundamental vibrations*. The natural coordinates are represented as a linear combination of the eigenvectors using the modal transformation; the modal coordinates are the corresponding weights or fractions:

$$\boldsymbol{\eta}_{\mathbf{q}} = \sum_i \bar{\boldsymbol{\eta}}_{\mathbf{q}_i}\xi_i\,. \tag{2.46}$$

As a solution of (2.45), it follows

$$\xi(t) = \cos(\Omega t)\,\xi_0 + \sin(\Omega t)\,\Omega^{-1}\dot{\xi}_0\,. \tag{2.47}$$

Comparing this with the solution for $\boldsymbol{\eta}_{\mathbf{q}}(t)$ gives:

$$\xi_0 = \mathbf{V}^{-1}\boldsymbol{\eta}_{\mathbf{q}0}\,, \tag{2.48}$$

$$\dot{\xi}_0 = \mathbf{V}^{-1}\dot{\boldsymbol{\eta}}_{\mathbf{q}0}\,. \tag{2.49}$$

For the derivation of the solution method, we have assumed that there are enough linearly independent eigenvectors in the system. This is not always correct, but it can be shown for conservative systems (2.45).

We consider the conservative system

$$\mathbf{M}\ddot{\boldsymbol{\eta}}_{\mathbf{q}} + \mathbf{K}\boldsymbol{\eta}_{\mathbf{q}} = \mathbf{0} \tag{2.50}$$

with symmetric and positive definite mass matrix $\mathbf{M}$ and symmetric stiffness matrix $\mathbf{K}$. Then, it should first be noted that $\mathbf{M}^{-1}\mathbf{K}$ is diagonalizable. In each case, there is a matrix $\mathbf{V}$, such that $\mathbf{V}^{-1}\mathbf{M}^{-1}\mathbf{K}\mathbf{V}$ can be decomposed into JORDAN *blocks* (Section 2.3.2.2). Without restriction, we can consider a single JORDAN block. Then column-by-column,

$$\mathbf{M}^{-1}\mathbf{K}\bar{\eta}_{\mathbf{q}_1} = \lambda \bar{\eta}_{\mathbf{q}_1} , \tag{2.51}$$

$$\mathbf{M}^{-1}\mathbf{K}\bar{\eta}_{\mathbf{q}_2} = \delta_1 \bar{\eta}_{\mathbf{q}_1} + \lambda \bar{\eta}_{\mathbf{q}_2} , \tag{2.52}$$

$$\mathbf{M}^{-1}\mathbf{K}\bar{\eta}_{\mathbf{q}_3} = \delta_2 \bar{\eta}_{\mathbf{q}_2} + \lambda \bar{\eta}_{\mathbf{q}_3} , \tag{2.53}$$

$$\vdots$$

With (2.51), it follows

$$\bar{\eta}_{\mathbf{q}_2}^T \mathbf{K}\bar{\eta}_{\mathbf{q}_1} = \lambda \bar{\eta}_{\mathbf{q}_2}^T \mathbf{M}\bar{\eta}_{\mathbf{q}_1} , \tag{2.54}$$

and with (2.52), it is

$$\bar{\eta}_{\mathbf{q}_1}^T \mathbf{K}\bar{\eta}_{\mathbf{q}_2} = \delta_1 \bar{\eta}_{\mathbf{q}_1}^T \mathbf{M}\bar{\eta}_{\mathbf{q}_1} + \lambda \bar{\eta}_{\mathbf{q}_1}^T \mathbf{M}\bar{\eta}_{\mathbf{q}_2} . \tag{2.55}$$

Finally, we obtain

$$\delta_1 = \frac{\bar{\eta}_{\mathbf{q}_1}^T \mathbf{K}\bar{\eta}_{\mathbf{q}_2} - \lambda \bar{\eta}_{\mathbf{q}_1}^T \mathbf{M}\bar{\eta}_{\mathbf{q}_2}}{\bar{\eta}_{\mathbf{q}_1}^T \mathbf{M}\bar{\eta}_{\mathbf{q}_1}} = 0 . \tag{2.56}$$

This procedure can be continued iteratively; so $\mathbf{M}^{-1}\mathbf{K}$ is diagonalizable and $\mathbf{V}$ contains the eigenvectors.

Because of the diagonalizability of $\mathbf{M}^{-1}\mathbf{K}$, the equations of motion (2.50) become decomposed into one-mass oscillators. What about $\mathbf{M}$ and $\mathbf{K}$? The example

$$\mathbf{M} = \mathbf{K} = \begin{pmatrix} 1 & 1 \\ 1 & 2 \end{pmatrix} \tag{2.57}$$

yields the double eigenvalue 1 for $\mathbf{M}^{-1}\mathbf{K}$; the eigenvectors are linearly independent but arbitrary. In particular, we can achieve that $\mathbf{V}^T\mathbf{M}\mathbf{V}$ and $\mathbf{V}^T\mathbf{K}\mathbf{V}$ are not diagonal. For simplicity, we assume that all *eigenvalues* of $\mathbf{M}^{-1}\mathbf{K}$ are different. Then, $\mathbf{V}^T\mathbf{M}\mathbf{V}$ and $\mathbf{V}^T\mathbf{K}\mathbf{V}$ are diagonal matrices, because apparently every eigenvalue-eigenvector pair $(\lambda, \bar{\eta}_{\mathbf{q}})$ of $\mathbf{M}^{-1}\mathbf{K}$ solves the equation

$$(\lambda \mathbf{M} - \mathbf{K})\, \bar{\eta}_{\mathbf{q}} = \mathbf{0} . \tag{2.58}$$

Thus, for two such pairs $(\lambda_n, \bar{\eta}_{\mathbf{q}_n})$ and $(\lambda_m, \bar{\eta}_{\mathbf{q}_m})$, it is

$$\bar{\eta}_{\mathbf{q}_n}^T \mathbf{M}\bar{\eta}_{\mathbf{q}_m} \lambda_m = \bar{\eta}_{\mathbf{q}_n}^T \mathbf{K}\bar{\eta}_{\mathbf{q}_m} , \tag{2.59}$$

$$\bar{\eta}_{\mathbf{q}_m}^T \mathbf{M}\bar{\eta}_{\mathbf{q}_n} \lambda_n = \bar{\eta}_{\mathbf{q}_m}^T \mathbf{K}\bar{\eta}_{\mathbf{q}_n} . \tag{2.60}$$

This yields

$$\bar{\eta}_{\mathbf{q}_n}^T \mathbf{M}\bar{\eta}_{\mathbf{q}_m} (\lambda_m - \lambda_n) = 0 \tag{2.61}$$

and finally our proposition. The work of the inertia and elastic forces of mode n with respect to the displacement of mode m vanishes. With $\eta_{\mathbf{q}} = \mathbf{V}\xi$, this means

$$\underbrace{\mathbf{V}^T\mathbf{M}\mathbf{V}}_{\mathbf{D}_M}\ddot{\xi} + \underbrace{\mathbf{V}^T\mathbf{K}\mathbf{V}}_{\mathbf{D}_K}\xi = \mathbf{0}^T \tag{2.62}$$

and therefore

$$\mathbf{V}^{-1}\mathbf{M}^{-1}\mathbf{K}\mathbf{V} = \mathbf{D}_M^{-1}\mathbf{D}_K\,. \tag{2.63}$$

With damping from the complex fundamental vibrations

$$\left\{\bar{\eta}_{\mathbf{q}_i} e^{-\delta_i t} e^{j\omega_i t}\right\}_{i=1}^{2f}, \tag{2.64}$$

we obtain real fundamental vibrations

$$\left\{e^{-\delta_i t}\mathbf{r}_i * \mathbf{e}^{j(\omega_i t\mathbf{E}+\varphi_i)}\right\}_{i=1}^{2f} \tag{2.65}$$

with componentwise multiplication $*$. Here, $\mathbf{r}$ is the real amplitude and φ is the phase shift, which are contained in the complex eigenvectors.

We consider a system with external forces $\mathbf{f}$:

$$\mathbf{M}\ddot{\eta}_{\mathbf{q}} + (\mathbf{D}+\mathbf{G})\,\dot{\eta}_{\mathbf{q}} + (\mathbf{K}+\mathbf{N})\,\eta_{\mathbf{q}} = \mathbf{f}\,. \tag{2.66}$$

Its solution consists of the solution of the homogeneous system (2.28) and a particular solution. We restrict ourselves to *periodic excitations*

$$\mathbf{f}(t) = \sum_{k=-\infty}^{\infty} \bar{\mathbf{F}}_k e^{jk\Omega t}\,, \quad \bar{\mathbf{F}}_k \in \mathbb{C}^n\,. \tag{2.67}$$

Since solutions of linear systems may be *superposed*, we consider, without restriction, even a *harmonic excitation*

$$\mathbf{f}(t) = \bar{\mathbf{F}}_k e^{jk\Omega t}\,, \tag{2.68}$$

which we intend to treat with a special approach for the particular solution:

$$\eta_{\mathbf{q}_{P_k}} = \bar{\eta}_{\mathbf{q}_{P_k}} e^{jk\Omega t}\,, \quad \bar{\eta}_{\mathbf{q}_{P_k}} \in \mathbb{C}^n\,. \tag{2.69}$$

Insertion into the differential equation yields

$$\bar{\eta}_{\mathbf{q}_{P_k}} = \underbrace{\left(-k^2\Omega^2\mathbf{M} + jk\Omega\mathbf{P} + \mathbf{R}\right)^{-1}}_{\mathrm{G}(j\Omega)}\bar{\mathbf{F}}_k\,. \tag{2.70}$$

The complex *frequency response function* $\mathrm{G}(j\Omega)$ contains amplitude and phase information. It gives the stationary transmission of the vibration system to a unit excitation function depending on the excitation angular frequency. The frequency response function is a frequency analysis tool, but can also be used to represent the particular solution of (2.69) in the time domain. Since in practice, the solution of the homogeneous system (2.28) decays because of damping, the steady-state component dominates after some time. This explains the nomenclature and the question, why usually only the stationary solution is determined if one is interested in the system's response to some excitation (Chapter 5).

General solution methods for multiple eigenvalues are discussed in the next section.

2.3.2 Linear First-Order Systems

A similar solution procedure as in the last section can be specified, if we start from the state space form at the end of Section 2.2:

$$\dot{\mathbf{x}} = \mathbf{A}\mathbf{x}\,, \quad \mathbf{x} \in \mathrm{I\!R}^{2f}\,, \quad \mathbf{A} \in \mathrm{I\!R}^{2f,2f}\,. \tag{2.71}$$

The matrix $\mathbf{A}$ is assumed to be constant. With the approach

$$\mathbf{x}(t) = \bar{\mathbf{x}} e^{\lambda t}\,, \tag{2.72}$$

we obtain a homogeneous linear system of equation for the eigenvectors $\bar{\mathbf{x}}$:

$$(\lambda \mathbf{E} - \mathbf{A})\,\bar{\mathbf{x}} = \mathbf{0}\,. \tag{2.73}$$

The characteristic equation for the eigenvalues λ is given via the characteristic polynomial:

$$P(\lambda) = \det(\lambda \mathbf{E} - \mathbf{A}) = 0\,. \tag{2.74}$$

For a real matrix $\mathbf{A}$, this polynomial is a real polynomial of degree $2f$. The $2f$ (complex) zeros λ_i are called eigenvalues.

2.3.2.1 Modal Behavior for a System without Multiple Eigenvalues

The eigenvector $\bar{\mathbf{x}}_i$ corresponding to the eigenvalue λ_i is determined by the homogeneous system of equations

$$(\lambda_i \mathbf{E} - \mathbf{A})\,\bar{\mathbf{x}}_i = \mathbf{0} \tag{2.75}$$

up to an arbitrary constant factor. The solution can be represented as a linear combination of these eigenvectors for systems without multiple eigenvalues [76]:

$$\mathbf{x}(t) = \sum_{i=1}^{2f} c_i \bar{\mathbf{x}}_i e^{\lambda_i t} \tag{2.76}$$

with the initial condition

$$\mathbf{x}_0 = \mathbf{x}(0) = \sum_{i=1}^{2f} c_i \bar{\mathbf{x}}_i \,. \tag{2.77}$$

The eigenvectors $\bar{\mathbf{x}}_i$ from (2.76) can be summarized to a modal matrix

$$\mathbf{X} = \left(\bar{\mathbf{x}}_1 \ \bar{\mathbf{x}}_2 \ldots \bar{\mathbf{x}}_{2f}\right) \,, \tag{2.78}$$

the eigenvalues λ_i to a diagonal matrix

$$\Lambda = \operatorname{diag}\{\lambda_i\} \tag{2.79}$$

and the constants c_i to the column vector

$$\mathbf{c} = \left(c_1, \cdots, c_{2f}\right)^T \,. \tag{2.80}$$

This yields

$$\mathbf{x}(t) = \sum_{i=1}^{2f} \bar{\mathbf{x}}_i e^{\lambda_i t} c_i = \left(\bar{\mathbf{x}}_1, \cdots, \bar{\mathbf{x}}_{2f}\right) \begin{pmatrix} e^{\lambda_1 t} & 0 & \cdots & 0 \\ 0 & \ddots & \ddots & \vdots \\ \vdots & \ddots & \ddots & 0 \\ 0 & \cdots & 0 & e^{\lambda_{2f} t} \end{pmatrix} \mathbf{c} = \mathbf{X} e^{\Lambda t} \mathbf{c} \,. \tag{2.81}$$

The above definition of $e^{\Lambda t}$ is an extension of the scalar calculation rules to matrices:

$$e^{\Lambda t} = \mathbf{E} + \Lambda t + \frac{1}{2}(\Lambda t)^2 + \text{hot} = \operatorname{diag}\left\{\sum_{k=0}^{\infty} \frac{(\lambda_i t)^k}{k!}\right\} = \operatorname{diag}\left\{e^{\lambda_i t}\right\} \,. \tag{2.82}$$

Further, it holds

$$\mathbf{x}_0 = \sum_{i=1}^{2f} \bar{\mathbf{x}}_i c_i = \mathbf{X}\mathbf{c} \quad \text{or} \quad \mathbf{c} = \mathbf{X}^{-1}\mathbf{x}_0 \,. \tag{2.83}$$

This yields the solution

$$\mathbf{x}(t) = \left(\mathbf{X} e^{\Lambda t} \mathbf{X}^{-1}\right) \mathbf{x}_0 \,. \tag{2.84}$$

The expression in brackets is called the *fundamental matrix*:

$$\Phi(t) = \mathbf{X} e^{\Lambda t} \mathbf{X}^{-1} \quad \text{with} \quad e^{\Lambda t} = \mathbf{X}^{-1} \Phi(t) \mathbf{X} \,. \tag{2.85}$$

$\Phi(t)$ and $e^{\Lambda t}$ are similar matrices.

Decoupled, the solution is particularly easy. This can be achieved by the modal transformation:

$$\mathbf{x}(t) = \mathbf{X}\zeta(t)\,, \tag{2.86}$$

$$\mathbf{x}(0) = \mathbf{X}\zeta(0)\,. \tag{2.87}$$

Then, it is

$$\dot{\zeta}(t) = \mathbf{X}^{-1}\mathbf{A}\mathbf{X}\zeta(t)\,.$$

If we now substitute the solution (2.84) in $\mathbf{0} = \dot{\mathbf{x}} - \mathbf{A}\mathbf{x}$, we obtain

$$\mathbf{0} = \mathbf{X}\Lambda e^{\Lambda t}\mathbf{X}^{-1}\mathbf{x}_0 - \mathbf{A}\left(\mathbf{X}e^{\Lambda t}\mathbf{X}^{-1}\right)\mathbf{x}_0 = [\mathbf{X}\Lambda - \mathbf{A}\mathbf{X}]\,e^{\Lambda t}\mathbf{X}^{-1}\mathbf{x}_0 \tag{2.88}$$

and thus the decoupled system

$$\dot{\zeta}(t) = \Lambda\zeta(t) \tag{2.89}$$

with the solution

$$\zeta = e^{\Lambda t}\zeta_0 \quad \text{and} \quad \zeta_0 = \mathbf{X}^{-1}\mathbf{x}_0\,. \tag{2.90}$$

This is known as the normal form of the vibration system and ζ as *normal coordinates.*

Normal coordinates ζ_i change over time independently of the other normal coordinates ζ_k. Normal coordinates ζ are complex in the general case and difficult to interpret. They result from a linear combination of position and velocity coordinates, and thus usually have no clear and plausible interpretation. However, because of the decoupling $\left(\dot{\zeta} = \Lambda\zeta\right)$, they offer mathematically significant benefits, particularly in the context of further considerations like the development of system models for controller design.

Each normal coordinate is either zero, an increasing/damped periodic oscillation, or an aperiodic motion. The properties result from the eigenvalues λ_i, which form complex conjugate pairs for matrices with real coefficients:

$$\lambda_i = \delta_i \pm j\omega_i \quad \text{with} \quad \{\delta, \omega\} \in \mathbb{R}\,. \tag{2.91}$$

With

$$e^{(\delta + j\omega)t} = e^{\delta t}\left(\cos\omega t + j\sin\omega t\right)\,,$$

the following basic characteristics result (Section 5.2):

1. $\delta_i = 0$
   According to Fig, 2.5a, one obtains a stationary continuous vibration.
2. $\omega_i = 0$
   Fig. 2.5b shows increasing or decaying *asymptotic solutions.*

3. $\delta_i \neq 0$ and $\omega_i \neq 0$
   One gets increasing and decaying oscillations (Fig. 2.5c and Fig. 2.5d).

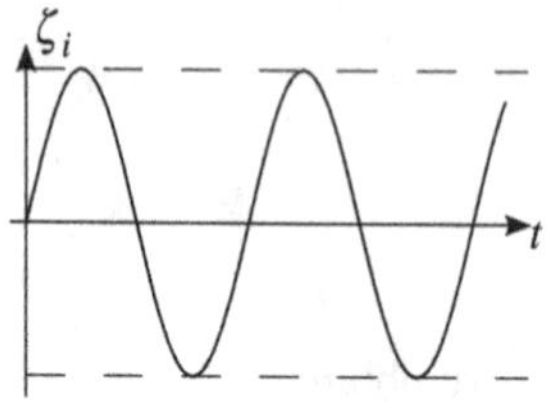

(a) Stationary continuous vibration.

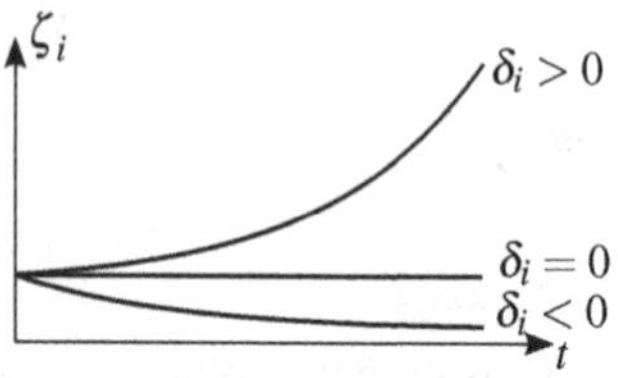

(b) Asymptotic solution.

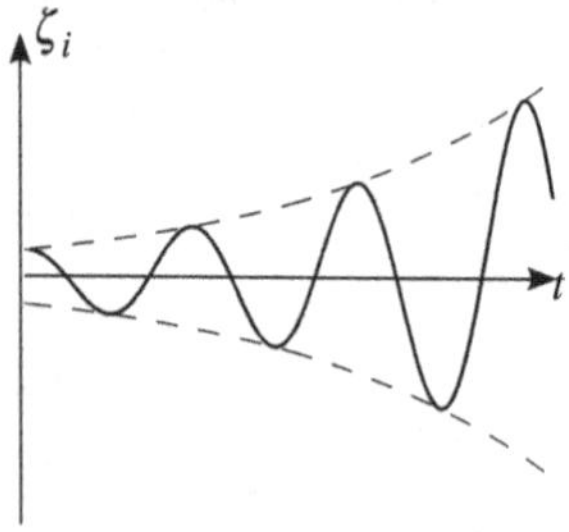

(c) Increasing oscillation.

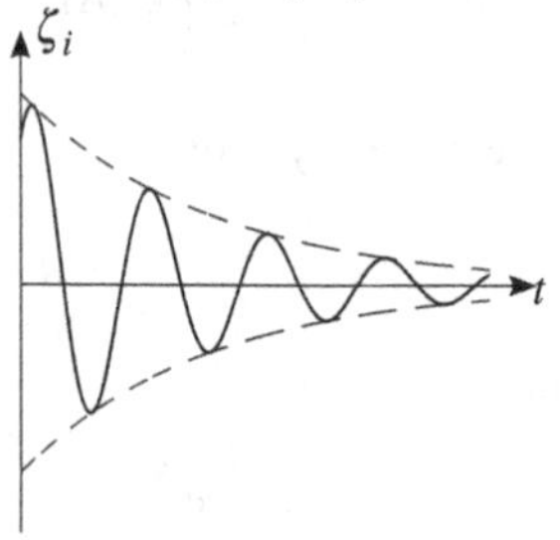

(d) Decaying oscillation.

Fig. 2.5. Basic characteristics.

Example 2.2 (Double pendulum). We consider a double pendulum with two point masses m and two massless pendulum rods of length l. As minimal coordinates we use the absolute angles $\eta_{\mathbf{q}} = (\varphi_1, \varphi_2)^T$. We assume that these are small: $\varphi_1 \ll 1$ and $\varphi_2 \ll 1$. The acceleration due to gravity with constant g acts in negative ${}_I\mathbf{y}$ direction. First, we derive the equations of motion using the LAGRANGE equations of the second kind. We linearize during this process in order to save computational effort. This procedure requires understanding of the system; linearization during and after setting up the equations of motion are not equivalent. Then, we calculate the mode shapes. We start with kinematic considerations and describe Cartesian positions and velocities of the two masses using the minimal coordinates:

- Cartesian positions:

$$\begin{aligned}
x_1 &= l \sin \varphi_1 \, , \\
y_1 &= -l \cos \varphi_1 \, , \\
x_2 &= l \sin \varphi_1 + l \sin \varphi_2 \, , \\
y_2 &= -l \cos \varphi_1 - l \cos \varphi_2 \, .
\end{aligned}$$

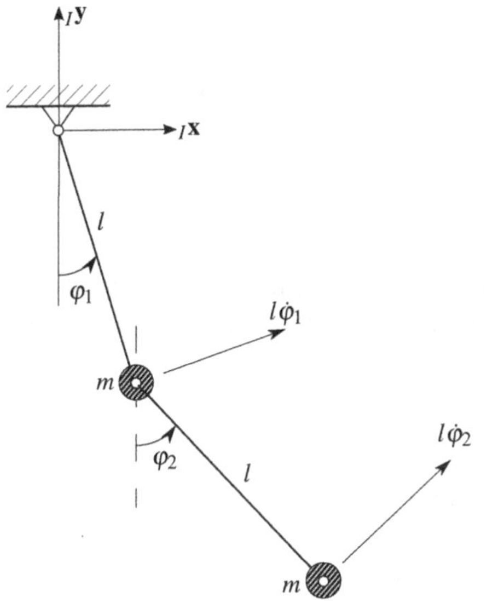

Fig. 2.6. Double pendulum.

- Cartesian velocities:

$$\begin{aligned}\dot{x}_1 &= l\dot{\varphi}_1 \cos\varphi_1 \ ,\\ \dot{y}_1 &= l\dot{\varphi}_1 \sin\varphi_1 \ ,\\ \dot{x}_2 &= l\dot{\varphi}_1 \cos\varphi_1 + l\dot{\varphi}_2 \cos\varphi_2 \ ,\\ \dot{y}_2 &= l\dot{\varphi}_1 \sin\varphi_1 + l\dot{\varphi}_2 \sin\varphi_2 \ .\end{aligned}$$

The square of the velocities

$$\begin{aligned}v_1^2 &= \left(\dot{x}_1^2 + \dot{y}_1^2\right) = (l\dot{\varphi}_1)^2 \ ,\\ v_2^2 &= \dot{x}_2^2 + \dot{y}_2^2 = (l\dot{\varphi}_1)^2 + (l\dot{\varphi}_2)^2 + 2l^2\dot{\varphi}_1\dot{\varphi}_2 \underbrace{(\cos\varphi_1\cos\varphi_2 + \sin\varphi_1\sin\varphi_2)}_{\cos(\varphi_2-\varphi_1)\doteq 1}\end{aligned}$$

is required for the calculation of the kinetic energy.

After having calculated the energies, we arrive at the evaluation of the LAGRANGE equations of the second kind.

- Energies ($\varphi_1 \ll 1$ and $\varphi_2 \ll 1$):

$$
\begin{aligned}
T &= \frac{1}{2}m(l\dot{\varphi}_1)^2 + \frac{1}{2}m\left[(l\dot{\varphi}_1)^2 + (l\dot{\varphi}_2)^2 + 2l^2\dot{\varphi}_1\dot{\varphi}_2\right] , \\
V &= mgl(1-\cos\varphi_1) + mgl\left[(1-\cos\varphi_1) + (1-\cos\varphi_2)\right] \\
&= mgl\left[\varphi_1^2 + \frac{1}{2}\varphi_2^2\right] + \text{hot} .
\end{aligned}
$$

  The potential energy is developed up to quadratic terms resulting in linear equations of motion after the evaluation of the LAGRANGE equations of the second kind.
- LAGRANGE equations of the second kind (1.151):
  With

$$
\begin{aligned}
\left(\frac{\partial T}{\partial \dot{\varphi}_1}\right) &= ml^2\left[2\dot{\varphi}_1 + \dot{\varphi}_2\right] , \\
\left(\frac{\partial T}{\partial \dot{\varphi}_2}\right) &= ml^2\left[\dot{\varphi}_2 + \dot{\varphi}_1\right] , \\
\left(\frac{\partial V}{\partial \varphi_1}\right) &= 2mgl\varphi_1 , \\
\left(\frac{\partial V}{\partial \varphi_2}\right) &= mgl\varphi_2 ,
\end{aligned}
$$

  we obtain

$$
ml^2\begin{pmatrix} 2 & 1 \\ 1 & 1 \end{pmatrix}\begin{pmatrix} \ddot{\varphi}_1 \\ \ddot{\varphi}_2 \end{pmatrix} + mgl\begin{pmatrix} 2 & 0 \\ 0 & 1 \end{pmatrix}\begin{pmatrix} \varphi_1 \\ \varphi_2 \end{pmatrix} = \begin{pmatrix} 0 \\ 0 \end{pmatrix} .
$$

  We summarize

$$
\mathbf{M}\ddot{\eta}_{\mathbf{q}} + \mathbf{K}\eta_{\mathbf{q}} = \mathbf{0}
$$

  with

$$
\mathbf{M} = \begin{pmatrix} 2 & 1 \\ 1 & 1 \end{pmatrix} , \quad \mathbf{K} = \left(\frac{g}{l}\right)\begin{pmatrix} 2 & 0 \\ 0 & 1 \end{pmatrix}
$$

  and

$$
\mathbf{M}^{-1} = \begin{pmatrix} 1 & -1 \\ -1 & 2 \end{pmatrix} .
$$

We condense the equation of motion taking advantage of $\omega = \sqrt{\frac{g}{l}}$

$$\ddot{\eta}_{\mathbf{q}} + \underbrace{\omega^2 \begin{pmatrix} 2 & -1 \\ -2 & 2 \end{pmatrix}}_{(\mathbf{M}^{-1}\mathbf{K})} \eta_{\mathbf{q}} = \mathbf{0}$$

and start the standard solution procedure:

- Characteristic equation with approach $\eta_{\mathbf{q}} = \bar{\eta}_{\mathbf{q}} e^{\lambda t}$:

$$0 = \det\left(\begin{pmatrix} \lambda^2 & 0 \\ 0 & \lambda^2 \end{pmatrix} + \omega^2 \begin{pmatrix} 2 & -1 \\ -2 & 2 \end{pmatrix}\right) = (2\omega^2 + \lambda^2)^2 - 2\omega^4 .$$

- Eigenvalues of the characteristic equation:

$$\lambda_{1,2,3,4} = \pm j\omega\sqrt{2 \mp \sqrt{2}} .$$

- Eigenvectors with a linear system of equations:

$$\begin{pmatrix} (2\omega^2 + \lambda_i^2) & -\omega^2 \\ -2\omega^2 & (2\omega^2 + \lambda_i^2) \end{pmatrix} \begin{pmatrix} \bar{\eta}_{\mathbf{q}_1} \\ \bar{\eta}_{\mathbf{q}_2} \end{pmatrix} = \mathbf{0} .$$

  From the first equation, we get the ratio

$$\frac{\bar{\eta}_{\mathbf{q}_{i2}}}{\bar{\eta}_{\mathbf{q}_{i1}}} = \left(\frac{2\omega^2 + \lambda_i^2}{\omega^2}\right) = \pm\sqrt{2}$$

  and finally

$$\bar{\eta}_{\mathbf{q}_1} = \begin{pmatrix} 1 \\ +\sqrt{2} \end{pmatrix} , \quad \bar{\eta}_{\mathbf{q}_2} = \begin{pmatrix} 1 \\ -\sqrt{2} \end{pmatrix} .$$

The eigenvectors are real and double. We summarize them to the modal matrix:

$$\mathbf{V} = \begin{pmatrix} 1 & 1 \\ +\sqrt{2} & -\sqrt{2} \end{pmatrix} .$$

Consequently, there are two eigen angular frequencies

$$\omega_1 = \omega\sqrt{2 - \sqrt{2}} = 0{,}765\sqrt{\frac{g}{l}} ,$$

$$\omega_2 = \omega\sqrt{2 + \sqrt{2}} = 1{,}848\sqrt{\frac{g}{l}} ,$$

which we arrange: $\cos(\Omega t) = \operatorname{diag}\{\cos\omega_i t\}$ and $\sin(\Omega t) = \operatorname{diag}\{\sin\omega_i t\}$. With this information, we consider again the solution:

- Solution with $\dot{\eta}_{\mathbf{q}_0} = \mathbf{0}$:

$$\eta_{\mathbf{q}} = \left[\mathbf{V}\cos(\Omega t)\,\mathbf{V}^{-1}\right]\eta_{\mathbf{q}_0}\,.$$

We obtain:

$$\begin{pmatrix}\varphi_1(t)\\ \varphi_2(t)\end{pmatrix} = \frac{1}{2}\begin{pmatrix}\left(\varphi_{10}+\frac{\varphi_{20}}{\sqrt{2}}\right)\cos(\omega_1 t)+\left(\varphi_{10}-\frac{\varphi_{20}}{\sqrt{2}}\right)\cos(\omega_2 t)\\ \sqrt{2}\left(\varphi_{10}+\frac{\varphi_{20}}{\sqrt{2}}\right)\cos(\omega_1 t)-\sqrt{2}\left(\varphi_{10}-\frac{\varphi_{20}}{\sqrt{2}}\right)\cos(\omega_2 t)\end{pmatrix}.$$

- Modal transformation:
  With the modal transformation

$$\eta_{\mathbf{q}} = \mathbf{V}\xi\,,$$

we have the opportunity for an interpretation. From

$$\xi_0 = \frac{1}{2}\begin{pmatrix}1 & \frac{1}{\sqrt{2}}\\ 1 & -\frac{1}{\sqrt{2}}\end{pmatrix}\begin{pmatrix}\varphi_{10}\\ \varphi_{20}\end{pmatrix},$$

it follows

$$\xi(t) = \frac{1}{2}\begin{pmatrix}\left(\varphi_{10}+\frac{\varphi_{20}}{\sqrt{2}}\right) & \cos(\omega_1 t)\\ \left(\varphi_{10}-\frac{\varphi_{20}}{\sqrt{2}}\right) & \cos(\omega_2 t)\end{pmatrix}.$$

- Interpretation:
  As a weighting function, modal coordinates describe the fraction of the mode shapes with respect to the overall motion. If we choose the coordinates of an eigenvector as the natural initial conditions, the other modal coordinate disappears. The two mode shapes can be realized easily in an experiment (Fig. 2.7).

The first mode (left) has the eigen angular frequency ω_1; the second mode (right) oscillates with the eigen angular frequency ω_2. The corresponding *eigenfrequencies* are defined as

$$f_k = \frac{\omega_k}{2\pi}\,.$$

The *period of oscillation* is obtained from

$$T_k = \frac{2\pi}{\omega_k}\,.$$

Since the ratio

$$\frac{\eta_{\mathbf{q}22}}{\eta_{\mathbf{q}21}} = -\sqrt{2}$$

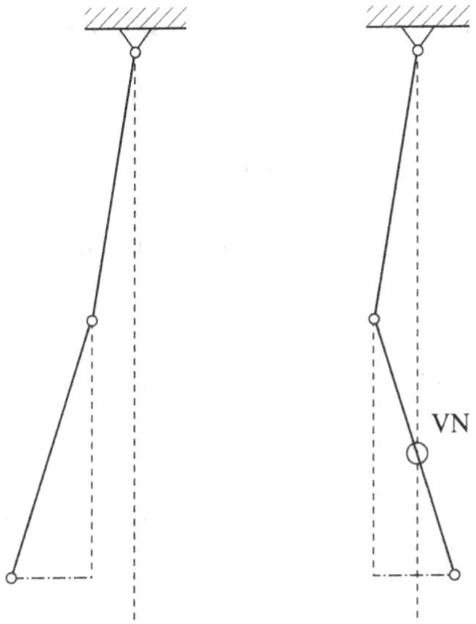

Fig. 2.7. Mode shapes of the double pendulum (VN = vibration node).

of the second mode is time independent, a so-called *vibration node* is formed. This is a characteristic feature of the mode shapes: the first mode does not have a vibration node, the second mode has one vibration node. In the illustration of Fig. 2.7 the argument only applies in a linearized sense for small deflections. A mode defines a synchronous response: amplitudes stay within a constant ratio.

For general damping according to (2.65), one obtains different amplitudes and phase shifts for the fundamental solution components. The relative behavior of the components is no longer time independent. Therefore, we will not find a vibration node in general damped systems.

Example 2.2 shows that the eigenvectors of $\mathbf{M}^{-1}\mathbf{K}$ are linearly independent but not orthogonal in general. Symmetry of $\mathbf{M}^{-1}\mathbf{K}$, however, is sufficient for the orthogonality of the eigenvectors. To see this, we combine two eigenvalue equations to the expression

$$0 = \eta_{\mathbf{q}_n}^T \mathbf{M}^{-1}\mathbf{K}\eta_{\mathbf{q}_m} - \eta_{\mathbf{q}_m}^T \mathbf{M}^{-1}\mathbf{K}\eta_{\mathbf{q}_n} = \eta_{\mathbf{q}_n}^T \eta_{\mathbf{q}_m} (\lambda_m - \lambda_n) \,.$$

For the restriction to simple eigenvalues, it follows the proposition: for multiple eigenvalues, there is freedom of choice.

If we consider the effect of the modal transformation on a right-hand side $\mathbf{f}$, then $\mathbf{V}^T\mathbf{f}$ defines its fraction in the equations for the modal coordinates ξ. There is no contribution to ξ_n, if and only if $\eta_{\mathbf{q}_n} \perp \mathbf{f}$. For excitation at point k with $\mathbf{f} = \mathbf{e}_k$, there is no contribution to ξ_n, if and only if

$$0 = \eta_{\mathbf{q}_n}^T \mathbf{e}_k = \eta_{\mathbf{q}_{n_k}}^T \,,$$

that is, if the discrete mode $\eta_{\mathbf{q}_n}$ has a vibration node at k.

2.3.2.2 Modal Behavior for a System with Multiple Eigenvalues

The system matrix **A** in (2.71) is diagonalizable with a similarity transformation only if enough linearly independent eigenvectors can be found. This can be guaranteed with pairwise distinct eigenvalues. If there are *multiple eigenvalues* λ_i as the zeros of the characteristic polynomial with the algebraic multiplicity ν_i , there exists a number d_i (geometric multiplicity) of linearly independent eigenvectors:

$$d_i = 2f - \mathrm{rg}\,(\lambda_i \mathbf{E} - \mathbf{A}) \tag{2.92}$$

where rg denotes the rank of the matrix. It holds [65, 46]

$$1 \leq d_i \leq \nu_i \quad \text{with} \quad i = 1, \cdots, m\,. \tag{2.93}$$

The matrix A is not diagonalizable, but a fundamental matrix $\Phi(t)$ can be derived. First, there always exists a regular matrix $\mathbf{X} \in \mathrm{I\!R}^{2f,2f}$, such that

$$\mathbf{X}^{-1}\mathbf{A}\mathbf{X} = \begin{pmatrix} \mathbf{J}_1 & & & \\ & \mathbf{J}_2 & \mathbf{0} & \\ & \mathbf{0} & \ddots & \\ & & & \mathbf{J}_m \end{pmatrix} =: \mathbf{J} \tag{2.94}$$

decomposes into so-called JORDAN *blocks*

$$\mathbf{J}_i := \begin{pmatrix} \lambda_i & 1 & & 0 \\ & \ddots & \ddots & \\ & & \ddots & 1 \\ 0 & & & \lambda_i \end{pmatrix}. \tag{2.95}$$

These differ from a diagonal matrix by the upper secondary diagonal which is filled with ones. The number of JORDAN blocks with the same eigenvalue λ_i is equal to the number of corresponding linearly independent eigenvectors. For each eigenvalue λ_i, there exist exactly d_i JORDAN blocks with the same λ_i on the diagonal according to (2.92).

With the transformation $\zeta = \mathbf{X}^{-1}\mathbf{x}$, we obtain

$$\dot{\zeta} = \mathbf{J}\zeta\,. \tag{2.96}$$

To derive the solution, we begin by considering the first JORDAN block of length l:

$$\begin{pmatrix} \dot{\zeta}_1 \\ \vdots \\ \dot{\zeta}_l \end{pmatrix} = \begin{pmatrix} \lambda_1 & 1 & & 0 \\ & \ddots & \ddots & \\ & & \ddots & 1 \\ 0 & & & \lambda_1 \end{pmatrix} \begin{pmatrix} \zeta_1 \\ \vdots \\ \zeta_l \end{pmatrix}. \tag{2.97}$$

The solution is obtained as a backward recursion of integration steps starting with the last equation. The initial step satisfies

$$\zeta_l = \zeta_{l,0} e^{\lambda_1 t} \,. \tag{2.98}$$

As

$$\dot{\zeta}_{l-1} = \lambda_1 \zeta_{l-1} + \zeta_l \Leftrightarrow \frac{\mathrm{d}\left(\zeta_{l-1} e^{-\lambda_1 t}\right)}{\mathrm{d}t} = \zeta_l e^{-\lambda_1 t} \,, \tag{2.99}$$

the recursion step reads

$$\zeta_{l-1} = \left(\zeta_{l,0} t + \zeta_{l-1,0}\right) e^{\lambda_1 t} \,. \tag{2.100}$$

Finally, we obtain

$$\begin{pmatrix} \zeta_1 \\ \vdots \\ \zeta_l \end{pmatrix} = \mathbf{K}_1(t) \begin{pmatrix} \zeta_{1,0} \\ \vdots \\ \zeta_{l,0} \end{pmatrix} \tag{2.101}$$

with

$$\mathbf{K}_1(t) := \begin{pmatrix} 1 & t & \frac{t^2}{2} & & \frac{t^{l-1}}{(l-1)!} \\ & & & \ddots & \\ & & \ddots & & \frac{t^2}{2} \\ & \ddots & & & t \\ 0 & & & & 1 \end{pmatrix} e^{\lambda_1 t} \,. \tag{2.102}$$

We repeat this with the other JORDAN blocks:

$$\zeta(t) = \underbrace{\begin{pmatrix} \mathbf{K}_1(t) & & \\ & & \mathbf{0} \\ & \ddots & \\ \mathbf{0} & & \\ & & \mathbf{K}_m(t) \end{pmatrix}}_{=:\mathbf{K}} \zeta_0 \,. \tag{2.103}$$

In natural coordinates, we get

$$\mathbf{x}(t) = \underbrace{\left(\mathbf{X}\mathbf{K}(t)\mathbf{X}^{-1}\right)}_{=:\boldsymbol{\Phi}(t)} \mathbf{x}_0 \,. \tag{2.104}$$

A method for calculating the columns of the matrix $\mathbf{X} = \left(\mathbf{x}_1, \cdots, \mathbf{x}_{2f}\right)$ is obtained directly from (2.94):

$$\mathbf{AX} = \mathbf{XJ} . \tag{2.105}$$

If we start again with the first JORDAN block $\mathbf{J}_1$ of length l, it follows

$$\begin{aligned}
\mathbf{A}\mathbf{x}_1 &= \lambda_1 \mathbf{x}_1 , \\
\mathbf{A}\mathbf{x}_2 &= \lambda_1 \mathbf{x}_2 + \mathbf{x}_1 , \\
\vdots &= \vdots \\
\mathbf{A}\mathbf{x}_l &= \lambda_1 \mathbf{x}_l + \mathbf{x}_{l-1} .
\end{aligned}$$

This iteration can be solved by forward recursion of systems of linear equations. The vector $\mathbf{x}_1$ is called eigenvector corresponding to the eigenvalue λ_1; the vectors $\mathbf{x}_2, \cdots, \mathbf{x}_l$ are called *generalized eigenvectors*. We proceed similarly with all other JORDAN blocks.

Example 2.3 (Wagon with a pendulum). A wagon (mass m_1) with a pendulum (mass m_2, length l) can move freely on horizontal rails (coordinate s). The pendulum can rotate without friction (angle φ). The acceleration of gravity acts in the negative ${}_I\mathbf{y}$ direction (Fig. 2.8).

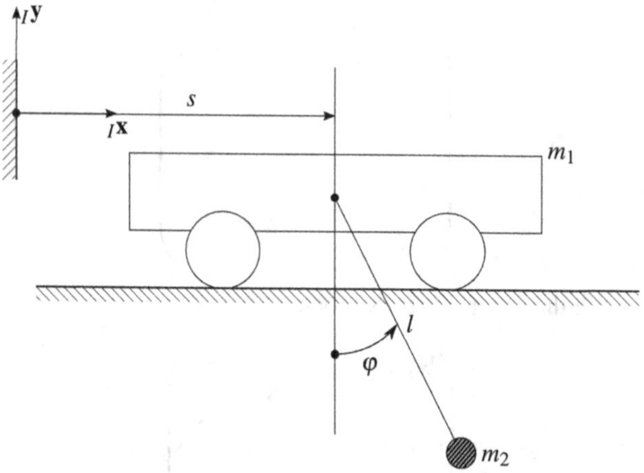

Fig. 2.8. Wagon with a pendulum.

We describe the motion with minimal coordinates $\mathbf{q} = (s, \varphi)^T$. The equations of motion are derived from energy expressions

$$\begin{aligned}
T &= \frac{1}{2} m_1 \dot{s}^2 + \frac{1}{2} m_2 \left(\dot{s}^2 + l^2 \dot{\varphi}^2 + 2l \cos\varphi \dot{s} \dot{\varphi} \right) , \\
V &= -m_2 g l \cos\varphi
\end{aligned}$$

with the LAGRANGE equations of the second kind:

$$\begin{pmatrix} m_1+m_2 & m_2 l\cos\varphi \\ m_2 l\cos\varphi & m_2 l^2 \end{pmatrix} \begin{pmatrix} \ddot{s} \\ \ddot{\varphi} \end{pmatrix} + \begin{pmatrix} -m_2 l\sin\varphi\dot{\varphi}^2 \\ m_2 g l\sin\varphi \end{pmatrix} = \begin{pmatrix} 0 \\ 0 \end{pmatrix} .$$

We are interested in small displacements and small velocities around the equilibrium position $\mathbf{q}_0 = \mathbf{0}$. Then, we have

$$\begin{pmatrix} m_1+m_2 & m_2 l \\ m_2 l & m_2 l^2 \end{pmatrix} \begin{pmatrix} \ddot{s} \\ \ddot{\varphi} \end{pmatrix} + \begin{pmatrix} 0 & 0 \\ 0 & m_2 g l \end{pmatrix} \begin{pmatrix} s \\ \varphi \end{pmatrix} = \begin{pmatrix} 0 \\ 0 \end{pmatrix}$$

with

$$(s,\varphi) = (s_0,\varphi_0) + (\eta_s,\eta_\varphi) .$$

For $m_1 = m_2 =: m$, the matrix $\mathbf{A}$ writes

$$\mathbf{A} = \begin{pmatrix} 0 & 0 & 1 & 0 \\ 0 & 0 & 0 & 1 \\ 0 & -l\omega & 0 & 0 \\ 0 & 2\omega & 0 & 0 \end{pmatrix}$$

with $\omega^2 = \frac{g}{l}$. This yields the following eigenvalues:

$$\lambda_{1,2} = \pm j\sqrt{2}\omega ,$$
$$\lambda_{3,4} = 0 .$$

The eigenvalue $\lambda_3 = 0$ has algebraic multiplicity $\nu_3 = 2$. We have to consider its geometric multiplicity separately:

$$d_3 = 2f - \mathrm{rg}(\lambda_3\mathbf{E} - \mathbf{A}) = 1 .$$

The matrix $\mathbf{A}$ cannot be diagonalized. Thus, for the corresponding JORDAN normal form, it follows

$$\mathbf{J} = \begin{pmatrix} \mathbf{J}_1 & & \\ & \mathbf{J}_2 & \\ & & \mathbf{J}_3 \end{pmatrix}$$

with

$$\mathbf{J}_1 = \lambda_1 \in \mathrm{IR}^{1,1} ,$$
$$\mathbf{J}_2 = \lambda_2 \in \mathrm{IR}^{1,1} ,$$
$$\mathbf{J}_3 = \begin{pmatrix} 0 & 1 \\ 0 & 0 \end{pmatrix} \in \mathrm{IR}^{2,2} .$$

The matrix $\mathbf{X}$, which transforms $\mathbf{A}$ to the JORDAN normal form according to (2.94), can be calculated with (2.105):

$$\mathbf{X} = \begin{pmatrix} l & l & 1 & 1 \\ -2 & -2 & 0 & 0 \\ j\sqrt{2}\omega l & -j\sqrt{2}\omega l & 0 & 1 \\ -j2\sqrt{2}\omega & j2\sqrt{2}\omega & 0 & 0 \end{pmatrix} .$$

The first three columns $\mathbf{x}_1$, $\mathbf{x}_2$, and $\mathbf{x}_3$ of $\mathbf{X}$ are the eigenvectors corresponding to the eigenvalues λ_1, λ_2, and λ_3. The fourth column is the generalized eigenvector for the double eigenvalue $\lambda_3 = \lambda_4 = 0$ (rigid body motion).

2.3.2.3 Forced Oscillations

We consider an externally excited system in the state space form of the equations (2.16) to (2.18):

$$\dot{\mathbf{x}}(t) = \mathbf{A}\mathbf{x}(t) + \mathbf{b}(t) . \tag{2.106}$$

The matrix $\mathbf{A}$ is assumed to be constant. The solution of (2.106) consists of a *homogeneous part* (2.84) for $\dot{\mathbf{x}} = \mathbf{A}\mathbf{x}$ and a *particular part* for the external excitation [46]:

$$\mathbf{x}(t) = \boldsymbol{\Phi}(t)\mathbf{x}_0 + \mathbf{x}_p(t) . \tag{2.107}$$

The particular solution $\mathbf{x}_p$ can be found by *variation of constants*:

$$\mathbf{x}_p(t) = \boldsymbol{\Phi}(t)\mathbf{c}(t) . \tag{2.108}$$

From (2.106), we obtain the relation

$$\dot{\boldsymbol{\Phi}}(t)\mathbf{c}(t) + \boldsymbol{\Phi}(t)\dot{\mathbf{c}}(t) = \dot{\mathbf{x}}_p(t) = \mathbf{A}\mathbf{x}_p(t) + \mathbf{b}(t) = \mathbf{A}\boldsymbol{\Phi}(t)\mathbf{c}(t) + \mathbf{b}(t) . \tag{2.109}$$

With the fundamental solution $\dot{\boldsymbol{\Phi}}(t) = \mathbf{A}\boldsymbol{\Phi}(t)$, it follows:

$$\boldsymbol{\Phi}(t)\dot{\mathbf{c}}(t) = \mathbf{b}(t) . \tag{2.110}$$

Due to (2.85), it holds

$$\dot{\mathbf{c}}(t) = \boldsymbol{\Phi}(-t)\mathbf{b}(t) , \tag{2.111}$$

which we integrate:

$$\mathbf{c}(t) = \mathbf{c}_0 + \int_0^t \boldsymbol{\Phi}(-\tau)\,\mathbf{b}(\tau)\,\mathrm{d}\tau . \tag{2.112}$$

Substituting the values $\mathbf{c}_0 = 0$ for $t = 0$ and noting, that $\boldsymbol{\Phi}(t)\boldsymbol{\Phi}(-\tau) = \boldsymbol{\Phi}(t-\tau)$ according to (2.85), yield the particular solution

$$\mathbf{x}_p = \underbrace{\int_0^t \boldsymbol{\Phi}(t-\tau)\,\mathbf{b}(\tau)\,\mathrm{d}\tau}_{\text{Duhamel integral}} \tag{2.113}$$

and the overall solution

$$\mathbf{x}(t) = \boldsymbol{\Phi}(t)\mathbf{x}_0 + \int_0^t \boldsymbol{\Phi}(t-\tau)\,\mathbf{b}(\tau)\,\mathrm{d}\tau\,. \tag{2.114}$$

Example 2.4 (Periodic excitation). Many technically important excitations in machine dynamics can be represented as *periodic oscillations*. As a representative, we consider the right-hand side

$$\mathbf{b}(t) = \mathbf{b}_0 e^{j\omega t}\,.$$

More complicated periodic excitation can be represented by FOURIER series; the solution is given by the *superposition principle* [41]. From (2.114), we get:

$$\mathbf{x}(t) = \boldsymbol{\Phi}(t)\mathbf{x}_0 + \int_0^t \boldsymbol{\Phi}(t-\tau)\,\mathbf{b}_0 e^{i\omega\tau}\mathrm{d}\tau\,.$$

Since according to (2.85), the fundamental matrix satisfies $\boldsymbol{\Phi}(t) = \mathbf{X}e^{\boldsymbol{\Lambda}t}\mathbf{X}^{-1}$, one has to evaluate some simple integrals with exponential functions in the equation above.

A common and relatively simple way to describe *forced oscillations* is via the LAPLACE *transformation* [76]. The function

$$F(s) = \mathscr{L}\{f\}(s) := \int_0^\infty e^{-st} f(t)\mathrm{d}t \tag{2.115}$$

is called LAPLACE transform of $f(t)$. The value s is a complex variable. Applying (2.115) to (2.106), we obtain

$$\mathbf{x}(s) = (s\mathbf{E} - \mathbf{A})^{-1}\,\mathbf{b}(s)\,. \tag{2.116}$$

These two rules are basically used:

$$\mathscr{L}\{a_1 f_1 + a_2 f_2\}(s) = a_1\mathscr{L}\{f_1\} + a_2\mathscr{L}\{f_2\}(s) \quad \text{(linearity)}\,, \tag{2.117}$$

$$\mathscr{L}\left\{f^{(n)}\right\}(s) = s^n\mathscr{L}\{f\}(s) - \sum_{k=0}^{n-1} f^{(k)}(0)s^{n-k-1} \quad \text{(differentiation pre-image)}\,. \tag{2.118}$$

Since we are only interested in the particular solution, the initial value is omitted in the differentiation rule. Equation (2.116) assumes the existence of the inverse $(s\mathbf{E} - \mathbf{A})^{-1}$. Since formally the inverse of a matrix is formed from its adjoint and its determinant [65], the determinant $\det(s\mathbf{E} - \mathbf{A})$ appears in the denominator of

(2.116). If the excitation frequencies contained in $\mathbf{b}(s)$ equal the eigenvalues of $\mathbf{A}$, it holds $\det(s\mathbf{E}-\mathbf{A})=0$ and we obtain resonances. The absolute value of $\mathbf{x}(s)$ defines the amplitude frequency response function; the phase of $\mathbf{x}(s)$ corresponds to the phase frequency response function. The evaluation of (2.116) is often easier to perform in individual cases than that of (2.114), because one obtains directly the amplitude and phase frequency response functions which are typically of interest (Section 2.3.1).

2.4 Stability of Linear Systems

We have already seen that the eigenvalues of the system play a central role for the time behavior. For the question of *stability*, we notice the following:

Theorems of LYAPUNOV [36, 45]
The equilibrium position $\dot{\mathbf{x}}(t)\equiv\mathbf{0}$ of the linear time-invariant system

$$\dot{\mathbf{x}}(t)=\mathbf{A}\mathbf{x}(t)\,,\quad \mathbf{x}(t_0)=\mathbf{x}_0 \qquad (2.119)$$

is

- *asymptotically stable*, if and only if all eigenvalues λ_i of $\mathbf{A}$ have negative real parts, $\Re(\lambda_i)<0$ (decaying oscillations),
- *stable*, if and only if $\mathbf{A}$ has no eigenvalues λ_i with positive real parts and for all eigenvalues with $\Re(\lambda_k)=0$, it holds: $\mathrm{Rg}(\lambda_k\mathbf{E}-\mathbf{A})=2f-\nu_k$,
- *unstable*, if and only if at least one eigenvalue λ_i of $\mathbf{A}$ has a positive real part, $\Re(\lambda_i)>0$, or a multiple eigenvalue with vanishing real part occurs, the algebraic multiplicity of which is larger than its geometric multiplicity.

2.4.1 *Criteria Based on the Characteristic Polynomial*

The eigenvalues λ_i of $\mathbf{A}$ follow from the characteristic equation

$$P(\lambda)=\det(\lambda\mathbf{E}-\mathbf{A})=a_0\lambda^{2f}+a_1\lambda^{2f-1}+\cdots+a_{2f-1}\lambda+a_{2f}=0\,. \qquad (2.120)$$

To assess the stability in terms of the eigenvalues, the solution of the characteristic equation is necessary. Of much greater interest is the question whether one can draw conclusions about the stability of the equilibrium position $\dot{\mathbf{x}}\equiv\mathbf{0}$ only by the structure of the characteristic polynomial. We give some criteria on the basis of the coefficients of the characteristic polynomial. Further details can be found in [38, 45, 10].

2.4.1.1 Stodola Criterion

Necessary for negative real parts of the eigenvalues λ_i is the condition

$$a_k > 0\,, \quad k = 0, 1, \cdots 2f\,. \tag{2.121}$$

Since the characteristic equation is determined only up to a factor, especially -1, the condition $a_0 > 0$ must be assumed without restriction.

With this criterion, we can exclude asymptotic stability. To prove asymptotic stability, the characteristic equation has to be further investigated.

2.4.1.2 Routh-Hurwitz Criterion

Necessary and sufficient for negative real parts of the eigenvalues λ_i is the condition

$$H_k > 0\,, \quad k = 1, \cdots, 2f\,. \tag{2.122}$$

The HURWITZ determinants H_k are leading principle minors of the HURWITZ matrix $\mathbf{H}$, which is formed from the coefficients of the characteristic polynomial:

$$\mathbf{H} = \begin{pmatrix} a_1 & a_3 & a_5 & a_7 & \cdots & 0 \\ a_0 & a_2 & a_4 & a_6 & \cdots & 0 \\ 0 & a_1 & a_3 & a_5 & \cdots & 0 \\ 0 & a_0 & a_2 & a_4 & \cdots & 0 \\ \cdots & \cdots & \cdots & \cdots & \cdots & 0 \\ 0 & \cdots & \cdots & \cdots & \cdots & a_n \end{pmatrix}\,. \tag{2.123}$$

The matrix $\mathbf{H} \in \mathrm{I\!R}^{2f,2f}$ is constructed as follows:

1. The first row is obtained by inserting the coefficients of the characteristic polynomial in the corresponding columns, where the index is increased by 2 from one column to the next.
2. In every further row, one makes use of the coefficients of the characteristic polynomial, whose index is lowered by 1 in comparison with the corresponding column of the preceding row. The row is completed according to the first rule.
3. As one leaves the domain of definition for the coefficients with this procedure, we have to set $a_k = 0$ for $k > 2f$ and $k < 0$.

Since the characteristic equation is again determined only up to a factor, especially -1, we have to assume $a_0 > 0$ without restriction.

Example 2.5 (ROUTH-HURWITZ criterion for $2f = 4$). We assume $a_0 > 0$ and take a look at

$$\mathbf{H} = \begin{pmatrix} a_1 & a_3 & 0 & 0 \\ a_0 & a_2 & a_4 & 0 \\ 0 & a_1 & a_3 & 0 \\ 0 & a_0 & a_2 & a_4 \end{pmatrix} .$$

Then, we get the following leading principal minors:

$$\begin{aligned} H_1 &= a_1 , \\ H_2 &= a_1 a_2 - a_0 a_3 , \\ H_3 &= a_1 a_2 a_3 - a_1^2 a_4 - a_0 a_3^2 , \\ H_4 &= a_4 H_3 . \end{aligned}$$

We draw the following conclusions from the ROUTH-HURWITZ criterion:

$$\begin{aligned} &H_4 > 0 \quad \Rightarrow \quad a_4 > 0 , \\ &H_3 = -a_1^2 a_4 + a_3 H_2 \quad \Rightarrow \quad a_3 > 0 , \\ &H_2 > 0 \quad \Rightarrow \quad a_2 > 0 , \\ &H_1 > 0 \quad \Rightarrow \quad a_1 > 0 . \end{aligned}$$

In particular, the STODOLA criterion follows:

$$a_0, a_1, a_2, a_3, a_4 > 0 .$$

If we assume the STODOLA criterion to be satisfied, then it is sufficient to analyze only every second HURWITZ determinant, that is either $H_1, H_3, H_5 \ldots > 0$ or $H_2, H_4, H_6 \ldots > 0$ (theorem of CREMER, see also LIÉNARD-CHIPART criterion).

2.4.1.3 Liénard-Chipart Criterion

The ROUTH-HURWITZ criterion includes the STODOLA criterion and can be combined with it as follows.

Necessary and sufficient for negative real parts of the eigenvalues λ_i is the condition

$$a_{2f} > 0,\ H_{2f-1} > 0,\ a_{2f-2} > 0,\ H_{2f-3} > 0, \cdots .$$

2.4.2 *Stability of Mechanical Systems*

For the stability analysis, we started from the theorems of LYAPUNOV (eigenvalue analysis) in a first step. For this, the explicit calculation of the eigenvalues is necessary. Consequently, we ask ourselves whether we can make statements about the stability without calculating the eigenvalues. The consideration of the characteristic equation yields the necessary coefficient criterion (STODOLA criterion). A further investigation gives the necessary and sufficient conditions of HURWITZ ("algebraic criterion"). Both criteria can be combined to the LIÉNARD-CHIPART criterion. For mechanical systems, all these criteria, however, do not make use of the special structure of the system matrix $\mathbf{A}$, that is from the physical meaning of the matrices $\mathbf{M}, \mathbf{D}, \mathbf{G}, \mathbf{K}, \mathbf{N}$. We discuss this briefly below [45].

2.4.2.1 Nongyroscopic Conservative Systems

The mechanical system

$$\mathbf{M}\ddot{\eta}_{\mathbf{q}} + \mathbf{K}\eta_{\mathbf{q}} = \mathbf{0} \tag{2.124}$$

is *critically stable*, that is stable but not asymptotically stable, if and only if the stiffness matrix $\mathbf{K}$ is positive definite:

$$\mathbf{K} = \mathbf{K}^T > 0\,. \tag{2.125}$$

For $\mathbf{K} < 0$ the system is unstable.

The scalar analogue for the one degree of freedom oscillator

$$m\ddot{\eta}_q + k\eta_q = 0 \quad \text{with} \quad \lambda_{1,2} = \pm j\sqrt{\frac{k}{m}} \tag{2.126}$$

is defined by the relations

$$k > 0 \rightarrow \text{critically stable}\,, \quad k < 0 \rightarrow \text{unstable}\,. \tag{2.127}$$

A matrix $\mathbf{K} = \mathbf{K}^T$ is called positive definite if the associated quadratic form satisfies $\mathbf{x}^T\mathbf{K}\mathbf{x} > 0$ for $\mathbf{x} \neq \mathbf{0}$. This is the case if and only if all leading principal minors are larger than zero or all eigenvalues are positive [65].

Since the potential energy of a conservative system can be written in the form

$$V = \frac{1}{2}\eta_{\mathbf{q}}^T\mathbf{K}\eta_{\mathbf{q}}\,, \tag{2.128}$$

the theorems of DIRICHLET and LAGRANGE follow directly from the matrix condition above.

Theorem of LAGRANGE (Mechanique Analytique, 1788):
If the potential energy V is a positive definite quadratic function in the neighborhood of an equilibrium position $\boldsymbol{\eta}_{\mathbf{q}} = \mathbf{0}$, then all eigenvalues λ_i^2 are negative real.

According to (2.128), it is $V(\mathbf{0}) = 0$ for $\boldsymbol{\eta}_{\mathbf{q}} = \mathbf{0}$. If $V > 0$ for $\boldsymbol{\eta}_{\mathbf{q}} \neq \mathbf{0}$, then $\mathbf{K}$ is positive definite. This means that $V(\mathbf{0})$ is an absolute minimum of $V(\boldsymbol{\eta}_{\mathbf{q}})$.

Theorem of DIRICHLET (1846):
If V has an absolute minimum in the equilibrium position $\boldsymbol{\eta}_{\mathbf{q}} = 0$, then it is stable.

If $\lambda^2 < 0$, then $\lambda = \pm j\omega$ is imaginary (ω real). As

$$e^{\lambda t} = e^{\pm j\omega t} = \cos(\omega t) \pm j \sin(\omega t) \ , \tag{2.129}$$

only critical stability can be concluded, so undamped vibrations about the equilibrium position occur.

2.4.2.2 Gyroscopic Conservative Systems

The mechanical system

$$\mathbf{M}\ddot{\boldsymbol{\eta}}_{\mathbf{q}} + \mathbf{G}\dot{\boldsymbol{\eta}}_{\mathbf{q}} + \mathbf{K}\boldsymbol{\eta}_{\mathbf{q}} = \mathbf{0} \tag{2.130}$$

is critically stable for $\mathbf{K} > 0$. For $\mathbf{K} < 0$, the system is stable if [40]

$$\det(\mathbf{G}) \neq 0 \text{ is sufficiently large} . \tag{2.131}$$

2.4.2.3 Damped Systems

The system

$$\mathbf{M}\ddot{\boldsymbol{\eta}}_{\mathbf{q}} + (\mathbf{D} + \mathbf{G})\,\dot{\boldsymbol{\eta}}_{\mathbf{q}} + \mathbf{K}\boldsymbol{\eta}_{\mathbf{q}} = \mathbf{0} \tag{2.132}$$

is asymptotically stable if $\mathbf{D} = \mathbf{D}^T > 0$ and the stiffness matrix is positive definite, regardless of $\mathbf{G}$.

Chapter 3
Linear Continuous Models

3.1 Models of Continuous Oscillators

Discrete systems are composed of rigid bodies, the essential property of which is that the distance between two points in the interior of such bodies remains constant with time. Elastic bodies are continua, which can deform elastically. We assume that their masses are homogeneous and isotropic. Furthermore, we restrict ourselves to linear-elastic bodies and thus to small deformations. The vibrations of these bodies are determined by their mass and stiffness distributions, similar to masses and springs in the discrete case. Each vibrating elastic system is characterized by eigenfrequencies and mode shapes. For each eigenfrequency, there is a corresponding mode shape which the structure takes when it oscillates at this frequency. There are infinite many eigenfrequencies and modes, usually in a systematic order. This property makes linear-elastic vibration systems relatively easy to understand, but it does not apply to all continuous systems, such as rotating fluids.

The calculation of eigenfrequencies and mode shapes is based on the equations of elastodynamics and can also be done analytically in simpler cases depending on the configuration of the analyzed component. In more complicated cases, we apply approximation methods. Continuum systems can be decomposed or discretized mathematically or physically. The *mathematical discretization methods* are based on the equations of motion in the form of *partial differential equations*, which are solved with sophisticated tools of numerical analysis (numerical integration, finite element method [76, 9, 64]). In a *physical discretization*, one decomposes the *continuum* into discrete elements, for example also into finite elements [74, 37, 78, 4], into element chains for the transfer-matrix method, and into a multibody system [11, 51, 62]. The discretizations often lead to similar systems of equations using different interpretations. Many of these discretization methods have the disadvantage that their complexity is very large and that the numerical results often lose physical transparency.

In practice, elastic vibration systems are frequently incorporated into discrete systems. Then, we have a vibration system with rigid and elastic components, which we call an *elastic multibody system* [11]. Since the motion should always be

F. Pfeiffer and T. Schindler, *Introduction to Dynamics*,
DOI: 10.1007/978-3-662-46721-3_3

described with a minimum effort and thus a minimum number of degrees of freedom, such a system is modelled sufficiently accurately with rigid body degrees of freedom for the rigid components and with elastic degrees of freedom for the elastic components; the crucial factor is the selection of the elastic degrees of freedom.

As we have already seen, the possible modes of vibration of an elastic body are given by its eigenmodes. Any mode of vibration resulting from excitation can be generated by superposition of eigenmodes. The contribution of each eigenmode to the total elastic deformation depends on the motion of the overall system, in which the elastic part is integrated. Therefore, we introduce *eigenmodes* together with a multiplicative time-dependent *modal coordinate* as *elastic degrees of freedom.*

The question of the necessary number of such degrees of freedom can be answered only approximately. It depends essentially on the effect of structural damping on the higher eigenfrequency amplitudes, on the type of excitation, and on the other frequencies expected in the system (related to other components or controllers). For example, operating frequencies and speeds are both limitations. In the case of elastic eigenfrequencies in such an excitation domain, they have to be considered. The modeling described above allows in any case the reduction of the degrees of freedom to the necessary minimum. How to obtain the mode shapes of the individual elastic components is a secondary question. Depending on the component, they can be calculated analytically or numerically.

3.2 Simple Examples of Continuous Vibrations

In the following, we discuss some simple analytical examples.

3.2.1 *Beam as a Bending Vibrator*

We consider a beam as a bending vibrator as shown in Fig. 3.1 [1, 69]. The deflection of the beam at point x and time t is denoted by $w(x,t)$, and the inclination of the

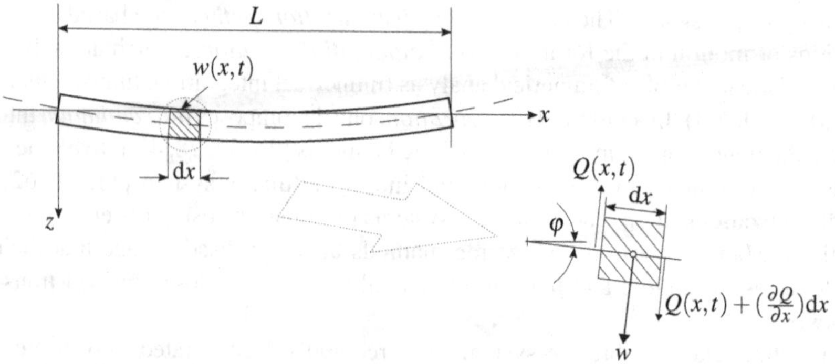

Fig. 3.1. Bending beam.

bending line by φ. Q is the transverse force, M the bending moment, EI the bending stiffness of the beam, and ρA its mass density [41]. The following relations hold:

- Kinematics:

$$\varphi \approx \left(\frac{\partial w}{\partial x}\right) . \tag{3.1}$$

- Elastostatics:

$$M(x) = -EI(x)\left(\frac{\partial^2 w}{\partial x^2}\right) , \tag{3.2}$$

$$Q(x) = \left(\frac{\partial M}{\partial x}\right) , \quad \frac{\partial Q}{\partial x} = -\frac{\partial^2}{\partial x^2}\left[EI(x)\frac{\partial^2 w}{\partial x^2}\right] . \tag{3.3}$$

- Momentum equation in z direction:

$$\rho A(x)\mathrm{d}x\left(\frac{\partial^2 w}{\partial t^2}\right) = -Q(x,t) + \left[Q(x,t) + \left(\frac{\partial Q}{\partial x}\right)\mathrm{d}x\right] = \left(\frac{\partial Q}{\partial x}\right)\mathrm{d}x . \tag{3.4}$$

This results in the equation of motion:

$$\frac{\partial^2}{\partial x^2}\left[EI(x)\left(\frac{\partial^2 w}{\partial x^2}\right)\right] + \rho A(x)\left(\frac{\partial^2 w}{\partial t^2}\right) = 0 . \tag{3.5}$$

In the case of a constant cross section and a constant bending stiffness, this yields the simplified equation

$$\left(\frac{\partial^4 w}{\partial x^4}\right) + \left(\frac{\rho A}{EI}\right)\left(\frac{\partial^2 w}{\partial t^2}\right) = 0 . \tag{3.6}$$

We introduce *separation of variables* according to BERNOULLI

$$w(x,t) = \sum_{i=1}^{\infty} w_i(x)q_i(t) = \mathbf{q}(t)^T \mathbf{w}(x) , \tag{3.7}$$

and then obtain for each individual summand

$$w_i^{(4)} q_i + \left(\frac{\rho A}{EI}\right) w_i \ddot{q}_i = 0 \tag{3.8}$$

and finally

$$\left(\frac{EI}{\rho A}\right)\frac{w_i^{(4)}}{w_i} = -\frac{\ddot{q}_i}{q_i} = \omega_i^2 . \tag{3.9}$$

The equality of the two ratios for all times and positions can only be achieved if they both take the same constant value ω_i^2. Therefore, we can split the partial differential equation into two ordinary differential equations

$$\ddot{q}_i + \omega_i^2 q_i = 0\,, \tag{3.10}$$
$$w_i^{(4)} - k_i^4 w_i = 0 \tag{3.11}$$

with the fundamental solutions

$$q_i(t) = a_i \cos(\omega_i t) + b_i \sin(\omega_i t)\,, \tag{3.12}$$
$$w_i(x) = A_i \cos(k_i x) + B_i \sin(k_i x) + C_i \cosh(k_i x) + D_i \sinh(k_i x) \tag{3.13}$$

and

$$k_i^4 = \left(\frac{\rho A}{EI}\right)\omega_i^2\,. \tag{3.14}$$

The four constants A_i, B_i, C_i, and D_i are determined from the *boundary conditions*. There are various possibilities, but we focus on the following cases:

- Clamped at both ends:

$$w(0,t) = w(L,t) = w'(0,t) = w'(L,t) = 0\,. \tag{3.15}$$

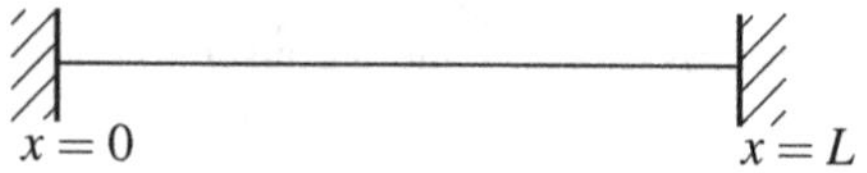

- Clamped at one end, simply supported at the other end:

$$w(0,t) = w'(0,t) = w(L,t) = 0\,, \tag{3.16}$$
$$M(L,t) = -EIw''(L,t) = 0\,. \tag{3.17}$$

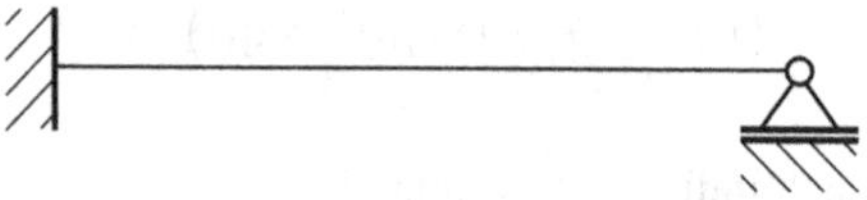

- Simply supported at both ends:

$$w(0,t) = w(L,t) = 0\,, \tag{3.18}$$
$$M(0,t) = -EIw''(0,t) = 0\,, \tag{3.19}$$
$$M(L,t) = -EIw''(L,t) = 0\,. \tag{3.20}$$

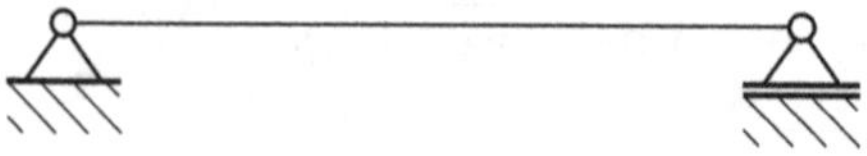

- Clamped at one end:

$$w(0,t) = w'(0,t) = 0\,, \tag{3.21}$$
$$M(L,t) = -EIw''(L,t) = 0\,, \tag{3.22}$$
$$Q(L,t) = -EIw'''(L,t) = 0\,. \tag{3.23}$$

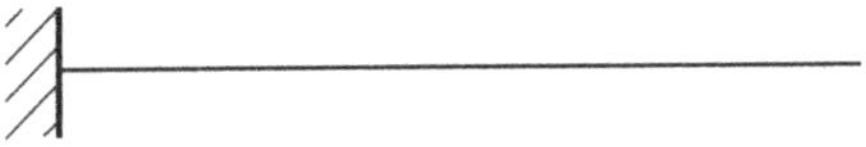

- Free at both ends:

$$M(0,t) = -EIw''(0,t) = 0\,, \tag{3.24}$$
$$Q(0,t) = -EIw'''(0,t) = 0\,, \tag{3.25}$$
$$M(L,t) = -EIw''(L,t) = 0\,, \tag{3.26}$$
$$Q(L,t) = -EIw'''(L,t) = 0\,. \tag{3.27}$$

For the four unknown constants A_i, B_i, C_i, and D_i, we get four determining equations. In the following we investigate the cantilevered case in more detail as an example. The *geometric boundary conditions* $w_i(0) = w_i'(0) = 0$ yield:

$$A_i + C_i = 0 \quad \text{and} \quad B_i + D_i = 0\,. \tag{3.28}$$

We obtain as a solution:

$$w_i(x) = A_i\left[\cos(k_i x) - \cosh(k_i x)\right] + B_i\left[\sin(k_i x) - \sinh(k_i x)\right]\,. \tag{3.29}$$

If we embed the *kinetic boundary conditions* $w_i''(L) = w_i'''(L) = 0$ into this solution, we get

$$\mathbf{0} = \begin{pmatrix} \cos(k_iL) + \cosh(k_iL) & \sin(k_iL) + \sinh(k_iL) \\ -\sin(k_iL) + \sinh(k_iL) & \cos(k_iL) + \cosh(k_iL) \end{pmatrix} \begin{pmatrix} A_i \\ B_i \end{pmatrix}\,. \tag{3.30}$$

This homogeneous system of equations for A_i and B_i only yields a nontrivial solution for a vanishing determinant. For (k_iL), the eigenvalue equation reads

$$\cos(k_iL)\cosh(k_iL) + 1 = 0\,. \tag{3.31}$$

There are infinitely many eigenvalues (Fig. 3.2), which can be approximated for large values (k_iL) as follows:

$$(k_iL) = (2i+1)\frac{\pi}{2}\,. \tag{3.32}$$

The corresponding eigen angular frequencies ω_i follow from (3.14). The corresponding mode shapes are sketched in Fig. 3.3. They are obtained by substituting

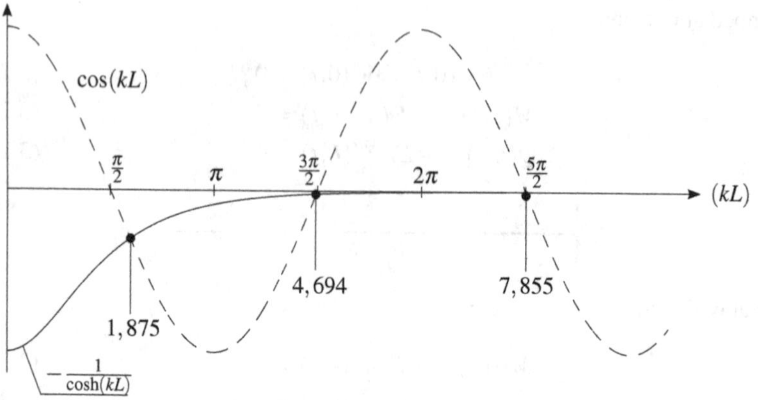

Fig. 3.2. Solution of the eigenvalue equation (3.31).

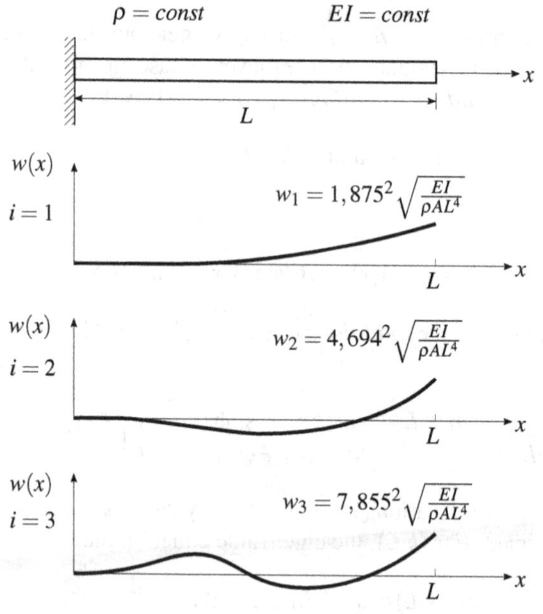

Fig. 3.3. Mode shapes of the bending beam.

(k_iL) in (3.30) and calculation of the coefficients A_i and B_i, which are finally inserted into (3.29):

$$w_i(x) = [\cos(k_iL) + \cosh(k_iL)]\,[\cos(k_ix) - \cosh(k_ix)] + [\sin(k_iL) - \sinh(k_iL)]\,[\sin(k_ix) - \sinh(k_ix)] \; . \tag{3.33}$$

The modes are orthogonal in the following sense (Section 3.3.1):

$$\int_0^L w_i(x)w_n(x)\mathrm{d}x = 0 \quad \text{for} \quad i \neq n \, , \tag{3.34}$$

$$\int_0^L w_i(x)w_n(x)\mathrm{d}x \neq 0 \quad \text{for} \quad i = n \, . \tag{3.35}$$

The solution for the beam vibration problem is obtained as the sum of all eigenmodes multiplied by the time-dependent coefficients $q_i(t)$:

$$w(x,t) = \sum_{i=1}^{\infty} q_i(t)w_i(x) = \mathbf{q}(t)^T\mathbf{w}(x) \; . \tag{3.36}$$

The coefficients a_i and b_i result from the *initial conditions* at time $t = 0$:

$$w_0(x) =: w(x,t=0) = \sum_{i=1}^{\infty} a_i w_i(x) \, , \tag{3.37}$$

$$\dot{w}_0(x) =: \dot{w}(x,t=0) = \sum_{i=1}^{\infty} b_i \omega_i w_i(x) \; . \tag{3.38}$$

Because of the orthogonality of the eigenmodes, multiplication by $w_i(x)$ and integrating over the beam length give

$$a_i = \frac{\int_0^L w_i(x)w_0(x)\mathrm{d}x}{\int_0^L w_i^2(x)\mathrm{d}x} \, , \tag{3.39}$$

$$b_i = \frac{\int_0^L w_i(x)\dot{w}_0(x)\mathrm{d}x}{\omega_i \int_0^L w_i^2(x)\mathrm{d}x} \; . \tag{3.40}$$

With (3.36), we see that we can incorporate an elastic component at any time as an element of a larger configuration without any change in the solution structure of the *initial boundary value problem.* In such a case, the values $q_i(t)$ are determined from the motion of the overall system. They weight the modes such that forced deformations occur.

3.2.2 *Beam as a Bending Vibrator with an End Mass*

The bending beam with an end mass (Fig. 3.4) is a technically important case, for example, for the modeling of wind turbines. Its description follows almost exactly

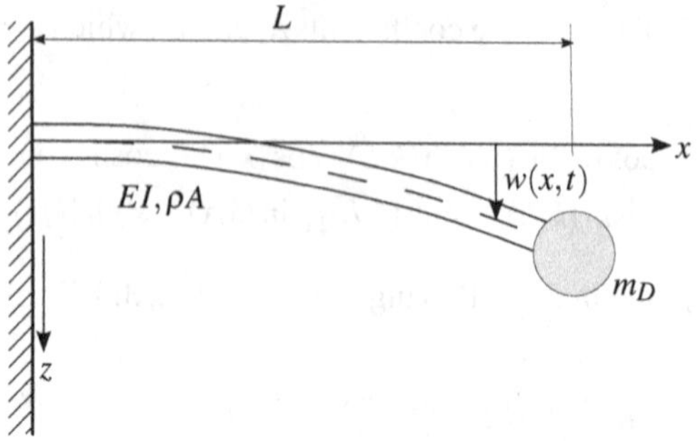

Fig. 3.4. Bending beam with an end mass.

the one without an end mass, since the end mass is noticeable only in the boundary conditions. We obtain the following relations:

- Equation of motion (3.5):

$$\frac{\partial^2}{\partial x^2}\left[EI(x)\frac{\partial^2 w}{\partial x^2}\right] = -\rho A(x)\frac{\partial^2 w}{\partial t^2} \,. \tag{3.41}$$

- Boundary conditions:

$$w(0,t) = w'(0,t) = 0 \,, \tag{3.42}$$

$$w''(L,t) = 0 \,, \tag{3.43}$$

$$\frac{\partial}{\partial x}\left[EI(x)\left(\frac{\partial^2 w}{\partial x^2}\right)\right]_L = m_D\left(\frac{\partial^2 w}{\partial t^2}\right)_L \,. \tag{3.44}$$

The inertia of the end mass must be compensated by the shear force.

We assume constant values EI and ρA along the beam. With separation of variables according to BERNOULLI

$$w(x,t) = \sum_{i=1}^{\infty} w_i(x) q_i(t) = \mathbf{q}(t)^T \mathbf{w}(x) \,, \tag{3.45}$$

it follows

$$\ddot{q}_i + \omega_i^2 q_i = 0 \,, \tag{3.46}$$

$$w_i^{(4)} - k_i^4 w_i = 0 \tag{3.47}$$

with

$$k_i^4 = \left(\frac{\rho A}{EI}\right)\omega_i^2 \,. \tag{3.48}$$

For the solution, we choose the same approach as in (3.13):

$$q_i(t) = a_i \cos(\omega_i t) + b_i \sin(\omega_i t) \,, \tag{3.49}$$
$$w_i(x) = A_i \cos(k_i x) + B_i \sin(k_i x) + C_i \cosh(k_i x) + D_i \sinh(k_i x) \,. \tag{3.50}$$

For the evaluation of the boundary conditions

$$w_i(0) = w_i'(0) = 0 \,, \tag{3.51}$$
$$w_i''(L) = 0 \,, \tag{3.52}$$
$$EIw_i'''(L) = -\omega^2 m_D w_i(L) \,, \tag{3.53}$$

we have to calculate the spatial derivatives:

$$w_i'(x) = k_i \left\{-A_i \sin(k_i x) + B_i \cos(k_i x) + C_i \sinh(k_i x) + D_i \cosh(k_i x)\right\} \,, \tag{3.54}$$
$$w_i''(x) = k_i^2 \left\{-A_i \cos(k_i x) - B_i \sin(k_i x) + C_i \cosh(k_i x) + D_i \sinh(k_i x)\right\} \,, \tag{3.55}$$
$$w_i'''(x) = k_i^3 \left\{A_i \sin(k_i x) - B_i \cos(k_i x) + C_i \sinh(k_i x) + D_i \cosh(k_i x)\right\} \,. \tag{3.56}$$

From the boundary conditions, we obtain

$$w_i(0) = 0 \Rightarrow A_i + C_i = 0 \,, \tag{3.57}$$
$$w_i'(0) = 0 \Rightarrow B_i + D_i = 0 \,, \tag{3.58}$$
$$w_i''(L) = 0 \Rightarrow A_i\left[-\cos(k_i L) - \cosh(k_i L)\right] + B_i\left[-\sin(k_i L) - \sinh(k_i L)\right] = 0 \,, \tag{3.59}$$

thus

$$A_i = -\left[\sin(k_i L) + \sinh(k_i L)\right] \,, \tag{3.60}$$
$$B_i = \left[\cos(k_i L) + \cosh(k_i L)\right] \,. \tag{3.61}$$

The mode shapes can be represented as follows:

$$\begin{aligned} w_i(x) = &\left[\cos(k_i L) + \cosh(k_i L)\right]\left[\sin(k_i x) - \sinh(k_i x)\right] \\ &- \left[\sin(k_i L) + \sinh(k_i L)\right]\left[\cos(k_i x) - \cosh(k_i x)\right] \,. \end{aligned} \tag{3.62}$$

The eigenvalues $(k_i L)$ follow from the fourth boundary condition (3.53),

$$\begin{aligned} &k_i^3 \left\{-\left[\cos(k_i L) + \cosh(k_i L)\right]^2 - \left[\sin(k_i L) + \sinh(k_i L)\right]\left[\sin(k_i L) - \sinh(k_i L)\right]\right\} \\ &= -\left(\frac{\omega^2 m_D}{EI}\right)\left\{\left[\cos(k_i L) + \cosh(k_i L)\right]\left[\sin(k_i L) - \sinh(k_i L)\right]\right. \\ &\quad \left. - \left[\sin(k_i L) + \sinh(k_i L)\right]\left[\cos(k_i L) - \cosh(k_i L)\right]\right\} \,. \end{aligned} \tag{3.63}$$

With $\frac{\omega_i^2}{EI} = \frac{k_i^4}{\rho A}$, we get the eigenvalue equation:

$$0 = 1 + \cos(k_iL)\cosh(k_iL) - (k_iL)\left(\frac{m_D}{\rho AL}\right)[\sin(k_iL)\cosh(k_iL) - \cos(k_iL)\sinh(k_iL)] \,. \tag{3.64}$$

Given m_D, the eigenvalues k_i have to be determined from this relation numerically. For $m_D = 0$, we get the same result as the one without an end mass. As in the last section, the initial conditions complete the overall solution:

$$w(x,t) = \sum_{i=1}^{\infty} w_i(x)q_i(t) = \mathbf{q}(t)^T\mathbf{w}(x) \,. \tag{3.65}$$

If we consider in addition the rotational inertia Θ of the end mass, the boundary condition (3.43) has to be replaced by the moment of momentum equation

$$\Theta\ddot{w}'(L,t) = M(L,t) = -EIw''(L,t) \,. \tag{3.66}$$

The general procedure for the determination of the coefficients stays the same.

3.2.3 Beam as a Torsional Vibrator with an End Mass

In machinery configurations, beams play an important role as torsional elements. We consider such a case as shown in Fig. 3.5 [1, 43, 69]. The torsion angle depends on the location x and the time t, $\varphi = \varphi(x,t)$, $GI_p(x)$ is the torsional stiffness with the polar area moment of inertia I_p, $J_p = \rho I_p$ is the moment of inertia per unit length with the density ρ, m_D is an end mass, and Θ is the corresponding moment of inertia. We have the following equations:

- Elastostatics:

$$M_t(x,t) = GI_p(x)\left(\frac{\partial\varphi}{\partial x}\right) \,. \tag{3.67}$$

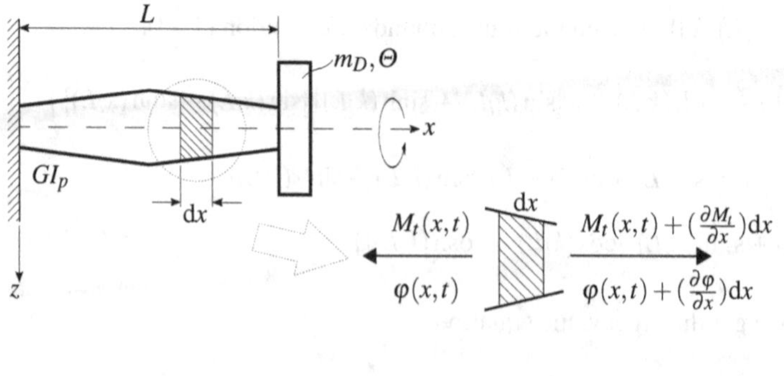

Fig. 3.5. Torsion element.

- Moment of momentum equation:

$$J_p(x)\mathrm{d}x\left(\frac{\partial^2\varphi}{\partial t^2}\right)=\left[M_t(x,t)+\left(\frac{\partial M_t}{\partial x}\right)\mathrm{d}x\right]-M_t(x,t)=\left(\frac{\partial M_t}{\partial x}\right)\mathrm{d}x\,. \quad (3.68)$$

This results in the equation of motion:

$$\frac{\partial}{\partial x}\left[GI_p(x)\left(\frac{\partial\varphi}{\partial x}\right)\right]-J_p(x)\left(\frac{\partial^2\varphi}{\partial t^2}\right)=0\,. \quad (3.69)$$

We assume constant values I_p, J_p along the torsion element and apply the separation of variables method according to BERNOULLI

$$\varphi(x,t)=\mathbf{q}(t)^T\varphi(x)=\sum_{i=1}^{\infty}\varphi_i(x)q_i(t)\,. \quad (3.70)$$

Two ordinary differential equations for $\varphi_i(x)$ and $q_i(t)$ result

$$\ddot{q}_i+\omega_i^2 q_i=0\,, \quad (3.71)$$

$$\varphi_i''+k_i^2\varphi_i=0 \quad (3.72)$$

with

$$k_i^2=\frac{J_p}{GI_p}\omega_i^2\,. \quad (3.73)$$

For $\varphi_i(x)$, the fundamental solution is

$$\varphi_i(x)=A_i\cos(k_ix)+B_i\sin(k_ix)\,. \quad (3.74)$$

The coefficients A_i and B_i are obtained from the boundary conditions:

$$\varphi(0,t)=0\,, \quad (3.75)$$

$$GI_p\left(\frac{\partial\varphi}{\partial x}\right)_L=-\Theta\left(\frac{\partial^2\varphi}{\partial t^2}\right)_L\,. \quad (3.76)$$

At the left end of the beam, no deformation is present, at the right end of the beam, the elastic torque is in balance with the inertia of the sheave. From $\varphi(0,t)=0$, it follows $A_i=0$ and therefore $\varphi_i(x)=\sin(k_ix)$, because the mode shapes are determined up to a factor. The boundary conditions at the right end of the beam result in the relationship

$$GI_pk_i\cos(k_iL)=\omega_i^2\Theta\sin(k_iL) \quad (3.77)$$

or after rewriting

$$k_i\left(\frac{GI_p}{\omega_i^2\Theta}\right) = k_i\left(\frac{GI_p}{\omega_i^2 J_p}\right)\left(\frac{J_p}{\Theta}\right) = \tan(k_i L)\ . \tag{3.78}$$

With k_i^2 from (3.73), an eigenvalue equation (Fig. 3.6) follows:

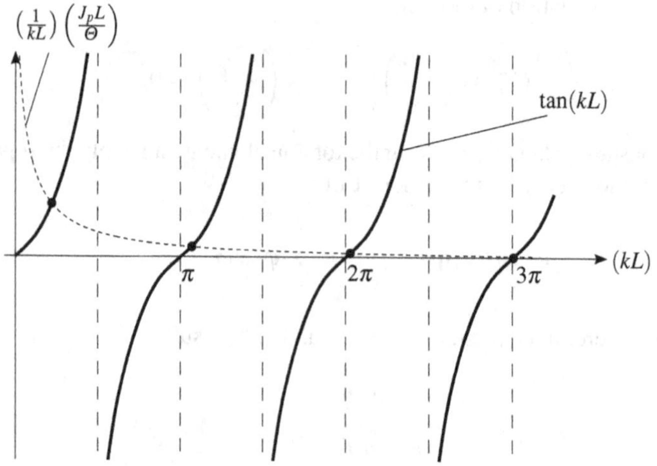

Fig. 3.6. Eigenvalue determination.

$$\tan(k_i L) = \left(\frac{1}{k_i L}\right)\left(\frac{J_p L}{\Theta}\right). \tag{3.79}$$

Taking into account the initial conditions, we get for the solution

$$\varphi(x,t) = \sum_{i=1}^{\infty} \sin(k_i x)\, q_i(t) = \mathbf{q}(t)^T \varphi(x)\ . \tag{3.80}$$

The first three modes are shown in Fig. 3.7.

For the case of a vanishing end mass ($\Theta = 0$), at the free end it is

$$GI_p\left(\frac{\partial\varphi}{\partial x}\right)_L = 0\,, \tag{3.81}$$

thus $\varphi'(L,t) = 0$. This yields $\cos(k_i L) = 0$ and

$$(k_i L) = (2i-1)\frac{\pi}{2}\ . \tag{3.82}$$

Considering the initial conditions, the solution satisfies

$$\varphi(x,t) = \sum_{i=1}^{\infty} \sin\left[(2i-1)\frac{\pi}{2}\left(\frac{x}{L}\right)\right] q_i(t)\ . \tag{3.83}$$

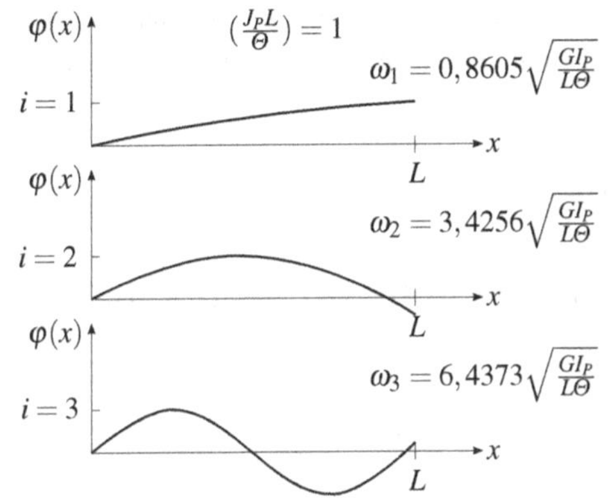

Fig. 3.7. Modes of the torsion vibration.

3.2.4 *Transverse Vibrations of a String*

The application range for string vibrations and for their model representation is very large, running from textile machines, belt drives in engines and continuously variable transmissions to high voltage cables for power transmission and carrier cables of cable cars.

By a string, we mean a one-dimensional elastic continuum, for example in the form of a thread, which is prestressed by a constant force (Fig. 3.8). The thread is clamped at $x = 0$ and at $x = L$. Oscillations are assumed to occur only in the direction transverse to the line continuum, they are small. Additional displacements in longitudinal direction due to the transverse vibrations are negligible. The vibration displacement is $w(x,t)$, and it depends on the location x and on the time t. The mass

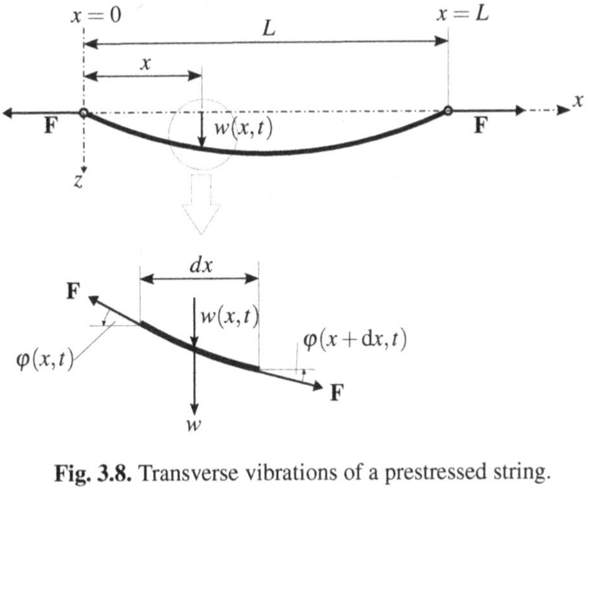

Fig. 3.8. Transverse vibrations of a prestressed string.

per unit length, also called line density, is constant $\mu = \rho A$ with density ρ and cross-sectional area A. Assuming small angles φ and applying the momentum theorem to a free piece of thread result in

$$\mu \mathrm{d}x(\frac{\partial^2 w}{\partial t^2}) = F\varphi(x+\mathrm{d}x,t) - F\varphi(x,t)\,. \tag{3.84}$$

It is $\varphi \approx (\frac{\partial w}{\partial x})$ and therefore $\varphi(x+\mathrm{d}x,t) - \varphi(x,t) \approx (\frac{\partial^2 w}{\partial x^2})\mathrm{d}x$. This yields the so-called *wave equation*:

$$\mu(\frac{\partial^2 w}{\partial t^2}) - F(\frac{\partial^2 w}{\partial x^2}) = 0 \quad \text{or} \quad (\frac{\partial^2 w}{\partial t^2}) - (\frac{F}{\mu})(\frac{\partial^2 w}{\partial x^2}) = 0\,. \tag{3.85}$$

This differential equation is valid for a large number of vibration phenomena in physics. The value $(\frac{F}{\mu}) = c^2$ is the square of the wave velocity. The standard form of the initial boundary value problem for the wave equation is therefore

$$\ddot{w}(x,t) - c^2 w''(x,t) = 0\,, \tag{3.86}$$
$$w(0,t) = 0\,, \quad w(L,t) = 0\,, \tag{3.87}$$
$$w(x,0) = w_0(x)\,, \quad \dot{w}(x,0) = \dot{w}_0(x)\,. \tag{3.88}$$

A fundamental solution is

$$w(x,t) = \sum_{i=1}^{\infty} w_i(x)\,[a_i \sin(\omega_i t) + b_i \cos(\omega_i t)] \tag{3.89}$$

with

$$w_i(x) = A_i \sin(\frac{\omega_i}{c}x) + B_i \cos(\frac{\omega_i}{c}x)\,. \tag{3.90}$$

From the boundary conditions, we get the eigenfunctions $w_i(x)$ and the eigenfrequencies ω_i:

$$w_i(x) = \sin\left(\pi i(\frac{x}{L})\right)\,, \tag{3.91}$$
$$\omega_i = \frac{\pi i}{L}\sqrt{\frac{F}{\mu}}\,. \tag{3.92}$$

The unknown values a_i and b_i can be determined from the initial conditions. For an infinitely long prestressed string, the wave equation (3.85) can be solved by the method of characteristics [76]. This corresponds to a left and a right moving wave with wave speed c.

3.3 Approximation of Continuous Vibration Systems

The vibrations of elastic continua often cannot be calculated analytically. In this case, we are looking for replacement systems that reflect the oscillatory behavior of the continuum with sufficient accuracy or which lead to an easier handling of the descriptive equations. The best known and most universal example is the *finite element method* (FEM), in which the continuum is replaced by a finite number of material elements (discretization). In this book, we do not address the FEM [12, 9]. Instead, we deal with two best practice methods for the treatment of continuous vibration systems, which moreover play a central role in the FEM: the methods of RAYLEIGH-RITZ and of BUBNOV-GALERKIN.

Both methods have been developed from mechanical problems. In his book [56] from 1877, RAYLEIGH described vibrations of elastic continua using a series of eigenmodes. RITZ abstracted the concept in 1908 [57] by formulating the equations as a variational problem, which he traced back (by approximating the integrand with mode shapes) to the minimization of a function with several variables. Independently in 1915, GALERKIN dealt with balance problems for thin plates. He approximated the functions appearing in the partial differential equations by suitable periodic functions, or as we know it today by the *method of weighted residuals*, and thus reduced the problem to the solution of algebraic equations [17]. According to [18] however, BUBNOV discovered the method of weighted residuals for the first time in 1913 for the approximation of partial differential equations. Today, we usually apply an additional integration by parts starting from the method of weighted residuals; based on this, we call the usage of finite-dimensional approximation spaces the BUBNOV-GALERKIN method (historically perhaps not totally correct). It is the starting point of the finite element method [12, 9]; in 1927, COURANT used *hat functions* as trial functions for the first time [13]. We start by providing some basic functional analytical background.

3.3.1 Function Systems and Completeness

Each scalar p-periodic function $f(x) = f(x+p)$ can be represented as a FOURIER *series* [76]. In practice, the series is often truncated after a finite number of elements; we obtain an approximation of f with the help of a *trigonometric polynomial*:

$$f(x) \approx f_N(x) = \sum_{i=-N}^{N} q_i e^{j\omega i x} \quad \text{with} \quad \omega = \frac{2\pi}{p}, \quad q_i = \frac{1}{p}\int_0^p f(x) e^{-j\omega i x} \mathrm{d}x . \tag{3.93}$$

The trigonometric polynomial is a linear combination of the linearly independent *trial functions* $w_i(x) = e^{j\omega i x}$. The trial functions are usually summarized in a so-called *function system* $\{w_i\}_i$. The set of all possible linear combinations, which also includes $f_N(x)$, forms a linear space, a so-called vector space. This can be extended by a *scalar product*: for elements of the function system, w_i and w_k, we define

$$< w_i, w_k >= \frac{1}{p} \int_0^p w_i(x) w_k^*(x) \mathrm{d}x \,, \tag{3.94}$$

with $w_k^*(x)$ being complex conjugate to $w_k(x)$. Substituting $w_k = w_i$, it follows

$$\| w_i \| = \sqrt{< w_i, w_i >} \tag{3.95}$$

and one obtains a *norm* for this linear space.

Convincing ourselves with the example of FOURIER series, it holds

$$< w_i, w_k >= \delta_{ik} \,. \tag{3.96}$$

With respect to the scalar product defined above, the basis functions w_i are *orthonormal*: they form an *orthonormal system*. The FOURIER coefficients satisfy

$$q_i =< f, w_i > \,. \tag{3.97}$$

They are obtained by minimizing the error

$$\Delta_N = \| f - f_N \|^2 \tag{3.98}$$

between the exact solution $f(x)$ and the approximate solution $f_N(x)$. For the minimum approximation error, we obtain

$$\Delta_{N,\min} = \| f \|^2 - \sum_{i=-N}^{N} q_i^2 \| w_i \|^2 \,. \tag{3.99}$$

The orthonormal system is called *complete*, if $\lim_{N\to\infty} \Delta_{N,\min} = 0$, that is if the so-called PARSEVAL theorem is satisfied

$$\| f \|^2 = \sum_{i=-\infty}^{\infty} q_i^2 \| w_i \|^2 \,. \tag{3.100}$$

This is the property of the representability of any scalar p-periodic function f with the help of the function system $\{w_i\}_i$ as FOURIER series.

For vector functions $f_l(x_1, \cdots, x_n)$ with $l \in \{1, \cdots, m\}$, the trial functions are linearly independent vector functions:

$$\mathbf{w}_i(\mathbf{x}) = \begin{pmatrix} w_{i,1}(\mathbf{x}) \\ \vdots \\ w_{i,m}(\mathbf{x}) \end{pmatrix} \quad \text{with} \quad \mathbf{x} = (x_1, \cdots, x_n)^T \,. \tag{3.101}$$

Concerning vibration problems, we usually have dependencies on $\mathbf{x}$ and t and a vector function $\mathbf{f} = \mathbf{f}(\mathbf{x}, t)$, that is $q_i = q_i(t)$. We met similar separation of variable approaches in Section 3.2.1; we now consider them for the calculation of approximations.

3.3.2 Rayleigh-Ritz Method

We consider linear elastic continua, the behavior of which can be described by the variational problem of HAMILTON's principle (1.174)[28]

$$\int_{t_1}^{t_2} (T - V)\,\mathrm{d}t \to \text{stationary}\,. \tag{3.102}$$

The expression

$$T = \frac{1}{2} \int_K \left(\dot{\mathbf{u}}^T \dot{\mathbf{u}}\right) \mathrm{d}m \tag{3.103}$$

is the kinetic energy of the continuum K,

$$\begin{aligned} V =& \frac{1}{2} \int_K \frac{1}{E} \left(\sigma_x^2 + \sigma_y^2 + \sigma_z^2\right) \mathrm{d}x\mathrm{d}y\mathrm{d}z \\ &- \frac{1}{2} \int_K \frac{2\mu}{E} \left(\sigma_x\sigma_y + \sigma_y\sigma_z + \sigma_z\sigma_x\right) \mathrm{d}x\mathrm{d}y\mathrm{d}z \\ &+ \frac{1}{2} \int_K \frac{1}{G} \left(\tau_{xy}^2 + \tau_{yz}^2 + \tau_{zx}^2\right) \mathrm{d}x\mathrm{d}y\mathrm{d}z \end{aligned} \tag{3.104}$$

is the potential energy of K (Fig. 3.9). Where E is *Young's modulus*, G is the *shear modulus*, μ is *Poisson's ratio*, and $\mathbf{u}(\mathbf{r},t)$ is a (linear-elastic) *displacement* of the position vector to the material point from its (stationary) reference position $\mathbf{r}$

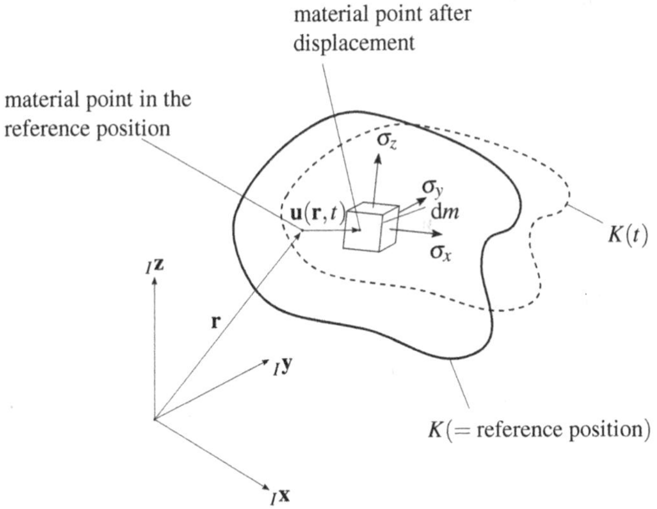

Fig. 3.9. Elastic deformation of a material point.

(LAGRANGE view). The stresses σ_x, σ_y σ_z, τ_{xy}, τ_{yz}, and τ_{zx} depend on the displacements $\mathbf{u}(\mathbf{r},t)$ due to the material law [12, 6, 69].

According to RITZ, we do not solve the variational problem (3.102) for the displacement $\mathbf{u}$ exactly. Instead, the displacement field $\mathbf{u}$ is approximated with a finite linear combination $\mathbf{u}_N$ of trial functions $\mathbf{w}_i(\mathbf{r})$ from a function system; therefore, the variational problem is reduced to a system of ordinary differential equations. With the coefficients $q_i(t)$, we approach $\mathbf{u}_N$ by

$$\mathbf{u}_N(\mathbf{r},t) = \sum_{i=1}^{N} \mathbf{w}_i(\mathbf{r})\, q_i(t) \,. \tag{3.105}$$

The trial functions $\mathbf{w}_i(\mathbf{r})$ are known. Then, T and V depend only on the coefficient functions $q_i(t)$, which will be determined such that the approximated functional (3.102) is stationary. This problem leads to LAGRANGE's equations of the second kind (1.151)

$$\frac{\mathrm{d}}{\mathrm{d}t}\left(\frac{\partial T}{\partial \dot{q}_i}\right) - \frac{\partial T}{\partial q_i} + \frac{\partial V}{\partial q_i} = 0 \tag{3.106}$$

for the coefficient functions q_i. As $\dot{\mathbf{u}}_N = \sum_i \mathbf{w}_i(\mathbf{r})\,\dot{q}_i(t)$, the kinetic energy T, which is approximated by the trial functions $\mathbf{w}_i$, only depends on $\dot{q}_i$. Then, (3.106) reduces to

$$\frac{\mathrm{d}}{\mathrm{d}t}\left(\frac{\partial T}{\partial \dot{q}_i}\right) + \frac{\partial V}{\partial q_i} = 0 \,. \tag{3.107}$$

We consider only continuous oscillations, where the displacement field $\mathbf{u}$ is planar, that is

$$\mathbf{u}(\mathbf{r},t) = (0,0,\tilde{w}(\mathbf{r},t))^T \,. \tag{3.108}$$

Then,

$$\tilde{w}_N(\mathbf{r},t) = \sum_{i=1}^{N} w_i(\mathbf{r})\, q_i(t) = \mathbf{q}(t)^T \mathbf{w}(\mathbf{r}) \tag{3.109}$$

with real-valued trial functions $w_i(\mathbf{r})$:

$$\mathbf{q}^T = (q_1, q_2, \cdots, q_N) \,, \tag{3.110}$$

$$\mathbf{w}^T = (w_1, w_2, \cdots, w_N) \,. \tag{3.111}$$

Furthermore,

$$T = \frac{1}{2}\dot{\mathbf{q}}^T \left(\int_K \mathbf{w}(\mathbf{r})\,\mathbf{w}(\mathbf{r})^T\,\mathrm{d}m\right)\dot{\mathbf{q}} = \frac{1}{2}\dot{\mathbf{q}}^T \mathbf{M}\dot{\mathbf{q}} \,, \tag{3.112}$$

which according to (3.107) leads to the following equations of motion –

RAYLEIGH-RITZ method:

$$\mathbf{M}\ddot{\mathbf{q}}+\left(\frac{\partial V}{\partial \mathbf{q}}\right)^T=\mathbf{0} \quad \text{with} \quad \mathbf{M}=\left(\int_K \mathbf{w}(\mathbf{r})\,\mathbf{w}(\mathbf{r})^T\,\mathrm{d}m\right). \tag{3.113}$$

The determination of the exact displacement field $\mathbf{u}(\mathbf{r},t)$ from the variational problem (3.102) is traced back to the determination of an approximate displacement field $\mathbf{u}_N(\mathbf{r},t)$, whose explicit representation is known with the solution of (3.113).

In the static case, it is $T=0$ and we obtain the equilibrium conditions

$$\left(\frac{\partial V}{\partial \mathbf{q}}\right)^T=\mathbf{0}\,. \tag{3.114}$$

The approximation method can be interpreted that among the infinitely many approximation systems, the RAYLEIGH-RITZ method picks the one, which satisfies the principle of d'ALEMBERT in the dynamic case (the LAGRANGE's equations) and the principle of virtual work (the equilibrium conditions) in the static case [77].

Example 3.1 (Cantilever beam according to RAYLEIGH-RITZ [69]). The bending vibrations of a cantilever beam are treated analytically in Section 3.2.1 (Fig. 3.10). We apply the RAYLEIGH-RITZ method. It is

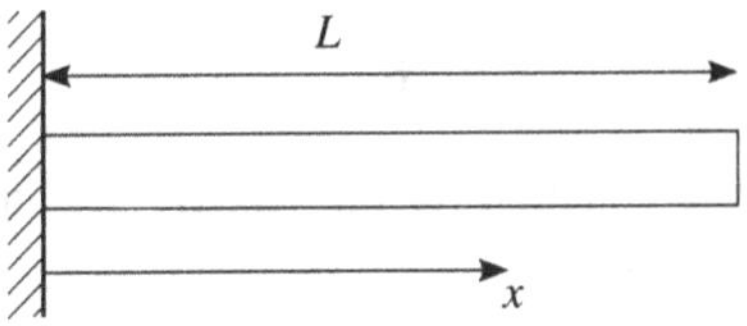

Fig. 3.10. Cantilever beam.

$$T=\frac{1}{2}\int_0^L \rho A\left(\frac{\partial w}{\partial t}\right)^2 \mathrm{d}x\,, \quad V=\frac{1}{2}\int_0^L EI\left(\frac{\partial^2 w}{\partial x^2}\right)^2 \mathrm{d}x\,.$$

We choose the approach

$$w(x,t)=\mathbf{q}^T(t)\mathbf{w}(x)\,.$$

The vectors $\mathbf{q}(t)$ and $\mathbf{w}(x)$ have only three components:

$$\mathbf{q}(t) = \begin{pmatrix} q_1(t) \\ q_2(t) \\ q_3(t) \end{pmatrix}, \quad \mathbf{w}(x) = \begin{pmatrix} \left(\frac{x}{L}\right)^2 \\ \left(\frac{x}{L}\right)^3 \\ \left(\frac{x}{L}\right)^4 \end{pmatrix}.$$

The trial functions represent the static bending line for a constant uniform load. From

$$V = \frac{1}{2}\mathbf{q}^T(t) \underbrace{\left\{ EI \int_0^L \mathbf{w}''(x)\mathbf{w}''^T(x)\mathrm{d}x \right\}}_{\mathbf{K}} \mathbf{q}(t),$$

it follows for the derivative

$$\left(\frac{\partial V}{\partial \mathbf{q}}\right)^T = \mathbf{K}\mathbf{q}.$$

This gives the equation of motion

$$\mathbf{M}\ddot{\mathbf{q}} + \mathbf{K}\mathbf{q} = \mathbf{0} \quad \text{with} \quad \mathbf{M} = \int_0^L \rho A \mathbf{w}\mathbf{w}^T \mathrm{d}x$$

and the matrices

$$\mathbf{M} = \begin{pmatrix} \frac{1}{5} & \frac{1}{6} & \frac{1}{7} \\ \frac{1}{6} & \frac{1}{7} & \frac{1}{8} \\ \frac{1}{7} & \frac{1}{8} & \frac{1}{9} \end{pmatrix} (\rho A L), \quad \mathbf{K} = \begin{pmatrix} 4 & 6 & 8 \\ 6 & 12 & 18 \\ 8 & 18 & \frac{144}{5} \end{pmatrix} \left(\frac{EI}{L^3}\right).$$

With the approach

$$\mathbf{q} = \bar{\mathbf{q}} e^{\lambda t},$$

we obtain the characteristic equation for the eigenvalues and the system of linear equations for the eigenvectors:

$$\det\left(\mathbf{M}\lambda^2 + \mathbf{K}\right) = 0, \quad \left(\mathbf{M}\lambda^2 + \mathbf{K}\right)\bar{\mathbf{q}} = \mathbf{0}.$$

We get the following eigenvalues:

	RAYLEIGH-RITZ	exact	rel. error
(k_1L)	1,876	1,875	0,05 %
(k_2L)	4,712	4,694	0,38 %
(k_3L)	19,261	7,855	145 %

The last line confirms the experience that one cannot determine the nth eigenvalue with n trial functions. The eigenvalues of the discrete system are larger than those of the continuous system, because the RAYLEIGH-RITZ method is consistently defined by an integral projection. The restriction to a finite number of trial functions acts as a constraint and yields a stiffening of the system. For a comparison of the approximated mode shapes

$$w_i(x,t) = e^{\lambda_i t} \bar{\mathbf{q}}_i^T \mathbf{w}(x)$$

with the exact mode shapes see Fig. 3.11.

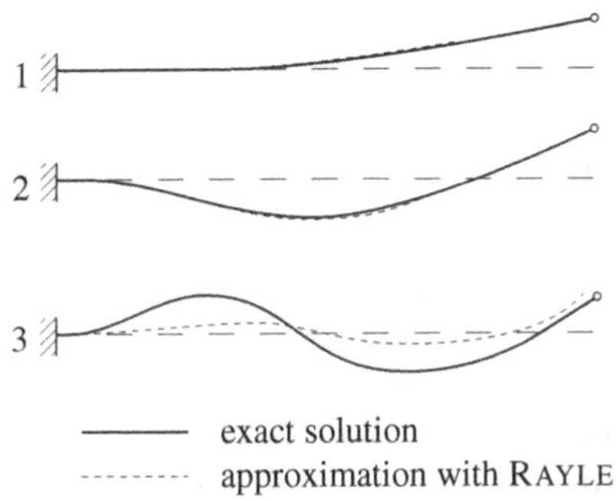

Fig. 3.11. Mode shapes (exact and approximation with RAYLEIGH-RITZ).

If we choose the mode shapes as trial functions in Example 3.1, we obtain a solution up to the corresponding order and diagonal mass and stiffness matrices. In general, we do not know the mode shapes. The trial functions have to be chosen arbitrarily, but with the boundary conditions to be satisfied (Section 3.3.4). We focus on global trial functions; for the finite element method, one selects local trial functions.

3.3.3 Bubnov-Galerkin Method

The models in mechanics lead to differential equations that represent balances of momentum and/or moment of momentum changes, that is sums of forces and/or moments. This also applies to continuum mechanics, where one has to deal with partial differential equations.

The approximation method of BUBNOV-GALERKIN does not primarily concern a variational problem, like (3.102), but the so-called *strong form*, that is a partial or an ordinary differential equation

$$D[\mathbf{u}] = \mathbf{0} \tag{3.115}$$

with the differential operator D, which we assume to be linear. We suppose that the displacement function $\mathbf{u}$ acts only in one plane and is thus defined by one component $\tilde{w}$ of $\mathbf{u}$.

Example 3.2 (Prestressed vibrating string: differential operator D). For the prestressed string (Fig. 3.12) with pretension force F defined in the reference state K, line density μ and deflection $w(\mathbf{r},t)$ at the location $\mathbf{r} = (x,0,0)^T$ and at the time t, we have obtained the wave equation as a descriptive differential equation in Section 3.2.4:

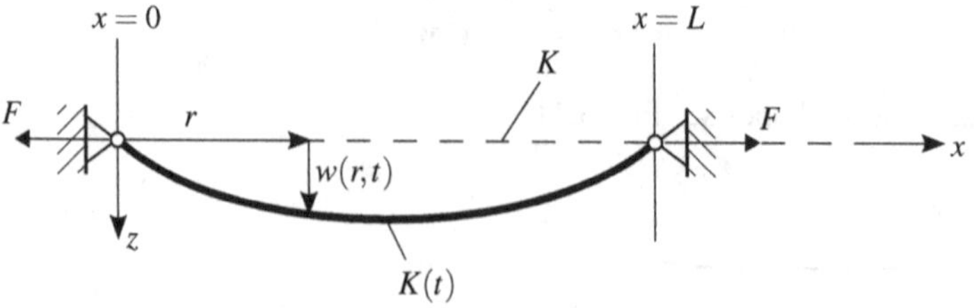

Fig. 3.12. Prestressed string.

$$\frac{\partial^2 w}{\partial x^2} - \frac{\mu}{F}\frac{\partial^2 w}{\partial t^2} = 0\,.$$

Since the position vector $\mathbf{r}$ represents a line continuum, the material points are defined by specifying only one coordinate, that is the *curve parameter* x for the reference configuration. The differential operator D is

$$D = \frac{\partial^2}{\partial x^2} - \frac{\mu}{F}\frac{\partial^2}{\partial t^2}\,.$$

We now assume that $\tilde{w}_N$ is an approximation for the exact solution $\tilde{w}$ of $D[\tilde{w}] = 0$ and use a function system with known trial functions w_i but with coefficient functions q_i to be determined:

$$\tilde{w}_N(x,t) = \sum_{i=1}^{N} w_i(x)q_i(t) = \mathbf{q}(t)^T\mathbf{w}(x)\,. \tag{3.116}$$

To obtain equations for the determination of the functions q_i, we require that the scalar product of the *residual*

$$r_N = D[\tilde{w}_N] - D[\tilde{w}] = D[\tilde{w}_N] \tag{3.117}$$

with given *weighting functions* g_k $(k = 1,\cdots,N)$ vanishes –

method of weighted residuals:

$$< r_N, g_k > = < D[\tilde{w}_N], g_k > = 0\,. \tag{3.118}$$

The scalar product $< \cdot,\cdot >$ is part of the function system of the trial functions (3.94). Up to a factor, equation (3.118) means to multiply the residual by the weighting functions and to integrate over the position variable.

If the weighting functions g_k belong to a complete function space, then we can represent $r = \lim_{N\to\infty} r_N$ itself as a series of the weighting functions g_k. Further,

$$< r, g_k >= 0 \quad \text{for} \quad (k = 1, 2, \cdots) \ , \tag{3.119}$$

that is r is orthogonal to every g_k. This implies $\lim_{N\to\infty} \| r_N \| = 0$. So, the residual r_N converges to 0 for $N \to \infty$.

If the approximation $\tilde{w}_N$ satisfies all the boundary conditions (Section 3.2), which are satisfied by the exact function $\tilde{w}$, then (under certain assumptions on the differential operator D [9]) from the convergence $r_N \to 0$ follows also the convergence $\tilde{w}_N \to \tilde{w}$:

$$\lim_{N\to\infty} \| \tilde{w}_N - \tilde{w} \| = 0 \, . \tag{3.120}$$

The method of weighted residuals is an intermediate step. Depending on how the weighting functions g_k are chosen, a number of methods can be derived. We focus on the classic BUBNOV-GALERKIN method and assume [76]

$$g_k = w_k \, . \tag{3.121}$$

The weighting functions correspond to the trial functions with which the solution $\tilde{w}$ is approximated. We split the differential operator into a time derivative component and a spatial derivative operator $K[\tilde{w}]$ and demonstrate the step to the BUBNOV-GALERKIN method with Example 3.2.

Example 3.3 (Prestressed vibrating string: BUBNOV-GALERKIN method). The differential operator D is composed of a time derivative component

$$-\frac{\mu}{F}\frac{\partial^2}{\partial t^2}$$

and a spatial derivative operator

$$K = \frac{\partial^2}{\partial x^2} \, .$$

The method of weighted residuals for $g_k = w_k$ is

$$\begin{aligned}
0 =& < D[\sum_{i=1}^{N} w_i(x) q_i(t)], w_k(x) >= \int_0^L D[\sum_{i=1}^{N} w_i(x) q_i(t)], w_k(x) \mathrm{d}x \\
=& \sum_{i=1}^{N} \int_0^L -\frac{\mu}{F} w_i(x) \frac{\partial^2 q_i(t)}{\partial t^2} w_k(x) + K[w_i(x)] q_i(t) w_k(x) \mathrm{d}x \\
=& -\frac{\mu}{F} \sum_{i=1}^{N} \int_0^L w_i(x) w_k(x) \mathrm{d}x \frac{\partial^2 q_i(t)}{\partial t^2} + \sum_{i=1}^{N} \int_0^L \frac{\partial^2 w_i(x)}{\partial x^2} w_k(x) \mathrm{d}x q_i(t) \, .
\end{aligned}$$

The trial and weighting functions w_i and w_k are known or chosen, such that the boundary conditions are satisfied for the system under consideration (Section 3.3.4). Then, the integrals can be calculated in advance and we obtain an ordinary differential equation for the time-dependent coefficients q_i in a similar way as for the RAYLEIGH-RITZ method

$$\mathbf{M}\ddot{\mathbf{q}} = \mathbf{K}\mathbf{q}$$

with mass and stiffness matrices

$$\mathbf{M}_{ki} = \frac{\mu}{F}\int_0^L w_i(x)w_k(x)\mathrm{d}x\,, \quad \mathbf{K}_{ki} = \int_0^L \frac{\partial^2 w_i(x)}{\partial x^2} w_k(x)\mathrm{d}x\,.$$

A drawback is the lack of *symmetry* of the stiffness matrix because of the second- and zeroth-order spatial derivatives of the trial or weighting functions. This is a major difference to the RAYLEIGH-RITZ method, which starts directly from the variational principle (3.102). We shift derivatives from the trial functions to the weighting functions by integration by parts until we get the desired symmetry:

$$\frac{\mu}{F}\sum_{i=1}^{N}\int_0^L w_i(x)w_k(x)\mathrm{d}x\frac{\partial^2 q_i(t)}{\partial t^2} = -\sum_{i=1}^{N}\left[\int_0^L \frac{\partial w_i(x)}{\partial x}\frac{\partial w_k(x)}{\partial x}\mathrm{d}x + \underbrace{\left.\frac{\partial w_i}{\partial x}w_k\right|_0^L}_{=0}\right] q_i(t)\,.$$

The *boundary integral* vanishes in our example, since the string is clamped on both sides and therefore the corresponding trial and weighting functions have to be chosen appropriately, we say they are *admissible*. We finally obtain the

BUBNOV-GALERKIN *method*

$$\mathbf{M}\ddot{\mathbf{q}} + \mathbf{K}\mathbf{q} = \mathbf{0}$$

with the mass and stiffness matrices

$$\mathbf{M}_{ki} = \frac{\mu}{F}\int_0^L w_i(x)w_k(x)\mathrm{d}x\,, \quad \mathbf{K}_{ki} = \int_0^L \frac{\partial w_i(x)}{\partial x}\frac{\partial w_k(x)}{\partial x}\mathrm{d}x$$

like in the RAYLEIGH-RITZ method.

Generally, the BUBNOV-GALERKIN method results from the method of weighted residuals by integration by parts, which is performed until the resulting differential operator has as a maximum of symmetry with respect to trial and weighting functions. The boundary integrals include so-called natural boundary conditions. We discuss the treatment of boundary conditions for all methods in the following section.

3.3.4 Boundary Conditions for the Rayleigh-Ritz and Bubnov-Galerkin Method

The solution of a variational problem or a partial differential equation must be supplemented by initial and boundary conditions. We characterize boundary conditions by an operator equation

$$\mathbf{R}[\tilde{w}] = \mathbf{0} \tag{3.122}$$

with

$$\mathbf{R} = (R_1, \cdots, R_m)^T \ . \tag{3.123}$$

The operator $\mathbf{R}$ contains as many suboperators as the problem has boundary conditions.

Example 3.4 (Cantilever beam: boundary conditions). By analogy with Example 3.14, we consider Fig. 3.13. We model geometric boundary conditions on the left and kinetic boundary conditions on the right:

$$\mathbf{R}[w] = \begin{pmatrix} w(0,t) \\ w'(0,t) \\ w''(L,t) + \frac{M}{EI} \\ w'''(L,t) + \frac{F}{EI} \end{pmatrix} = \mathbf{0} \quad \begin{matrix} \left.\vphantom{\begin{matrix} a \\ b \end{matrix}}\right\} \text{kinematic}\,, \\ \left.\vphantom{\begin{matrix} a \\ b \end{matrix}}\right\} \text{kinetic}\,. \end{matrix}$$

We denote boundary conditions, where only the boundary values of $\tilde{w}$ and $\tilde{w}'$ occur, as *geometric*, *kinematic*, or *essential* boundary conditions. Accordingly, we denote boundary conditions, where also the boundary values for $\tilde{w}''$, $\tilde{w}'''$ occur, as *free*, *kinetic*, or *natural* boundary conditions.

For the convergence $\tilde{w}_N \to \tilde{w}$ in the method of weighted residuals, we have the requirements in accordance with Section 3.3.3:

1. the N trial and weighting functions have to be from a complete function system,
2. all boundary conditions imposed on $\tilde{w}$ have to be satisfied also by w_k.

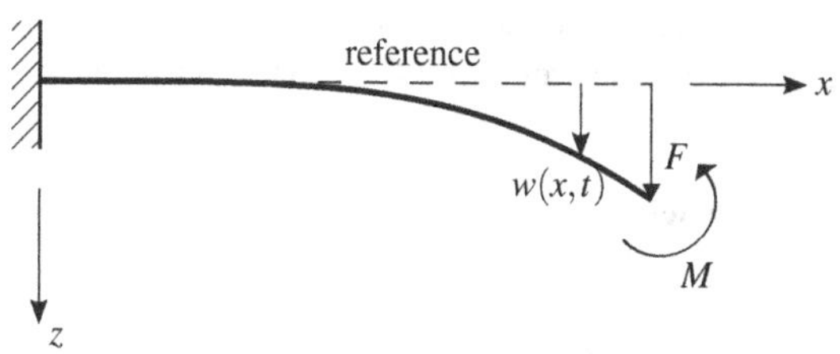

Fig. 3.13. Cantilever beam with constant end load and constant end torque.

Since the method of weighted residuals is based on the differential equation itself, the trial functions have to be admissible in the sense that they meet both the geometric and the free boundary conditions.

The BUBNOV-GALERKIN method is constructed from the method of weighted residuals by integration by parts. We choose $g_k = w_k$ with $g_k = 0$ and $g'_k = 0$ at geometric boundaries. Geometric boundary conditions vanish during the integration by parts, free boundary conditions are preserved naturally. So, $\tilde{w}_N$ converges to $\tilde{w}$, if

1. the N trial and weighting functions w_k are from a complete function system,
2. trial functions w_k satisfy the geometric boundary conditions.

The BUBNOV-GALERKIN method is based on the so-called *weak form* and the only requirement is fulfilling the geometric constraints regarding admissibility. The RAYLEIGH-RITZ method behaves accordingly. After defining the energy, for example, the corresponding potentials for the description of free boundary conditions, admissible trial functions must satisfy only the geometric boundary conditions.

Example 3.5 (Cantilever beam according to BUBNOV-GALERKIN [69]). In Example 3.1, we have treated the cantilever beam with the RAYLEIGH-RITZ method. Now we apply the BUBNOV-GALERKIN method. The differential equation is

$$D[w] = EI\left(\frac{\partial^4 w}{\partial x^4}\right) + \rho A\left(\frac{\partial^2 w}{\partial t^2}\right) = 0\,.$$

As in Example 3.1, the boundary conditions satisfy

$$\begin{aligned} w(0,t) &= w'(0,t) = 0\,, \\ w''(L,t) &= w'''(L,t) = 0\,. \end{aligned}$$

As an approach, we choose $w(x,t) = \mathbf{q}^T(t)\mathbf{w}(x)$ with the trial functions of the static bending line:

$$\mathbf{w}(x) = \begin{pmatrix} (\frac{x}{L})^2 \\ (\frac{x}{L})^3 \\ (\frac{x}{L})^4 \end{pmatrix}\,.$$

This approach satisfies the geometric boundary conditions. We insert it into the differential equation, multiply it by the weighting functions, and integrate over the length of the beam:

$$\int_0^L \left(EI\frac{\partial^4 \mathbf{w}^T}{\partial x^4}\mathbf{q} + \rho A\mathbf{w}^T\ddot{\mathbf{q}}\right) w_k \mathrm{d}x = 0\,.$$

Integration by parts with $w_k(0) = 0$ and $w'_k(0) = 0$ yields

$$\int_0^L \left(EIw_k'' \frac{\partial^2 \mathbf{w}^T}{\partial x^2} \mathbf{q} + \rho A w_k \mathbf{w}^T \ddot{\mathbf{q}} \right) \mathrm{d}x + \underbrace{\left(EIw_k \frac{\partial^3 \mathbf{w}^T}{\partial x^3} \mathbf{q} \right)_L}_{\text{force shifted by } w_k} - \underbrace{\left(EIw_k' \frac{\partial^2 \mathbf{w}^T}{\partial x^2} \mathbf{q} \right)_L}_{\text{moment rotated by } w_k'} = 0 \,.$$

The loads at the free boundary disappear in our example. Thus, we obtain the differential equation system of the RAYLEIGH-RITZ method

$$\mathbf{M}\ddot{\mathbf{q}} + \mathbf{K}\mathbf{q} = \mathbf{0}$$

from Example 3.1

3.3.5 *Choice of Trial Functions*

The efficiency of the approximation methods presented is highly dependent on the choice of the trial functions $\mathbf{w}$. Admissible trial functions must satisfy the geometric boundary conditions in each case. For the RAYLEIGH-RITZ or BUBNOV-GALERKIN method, the highest derivative degree is half the one occurring for the method of weighted residuals due to energy expressions or integration by parts. Admissible trial functions have to have only half differentiability. Because of the energy expressions or integration by parts, free boundary conditions are also included. These have to be provided for admissibility only in the method of weighted residuals.

Since we have limited ourselves to global trial functions, no further rules can be specified. We choose the trial functions to be as simple as possible. We try to ensure that the trial functions are qualitatively consistent with the expected mode shapes in reality, as we can expect satisfactory mathematical convergence only in this case. Frequently, it may be useful to choose the mode shapes of a replacement problem as trial functions, for which the vibration characteristics are close to the continuous component in the overall system. Such a replacement problem can be difficult to define (e.g., a rotating rod). In such cases, it might be better to choose trial functions from the physical intuition and adjust them iteratively.

In terms of automation and flexibility, finite element methods are preferable. They are based on the weak form and do not use global but local, mostly piecewise polynomial, interpolation functions. These can be locally refined and adapted automatically to load and stress curves [12, 9].

3.3.6 *Bending Vibrations of a Beam with Longitudinal Load*

We consider the bending vibrations of a beam with longitudinal load as shown in Fig. 3.14. Approximations are derived with the RAYLEIGH-RITZ and the BUBNOV-GALERKIN method.

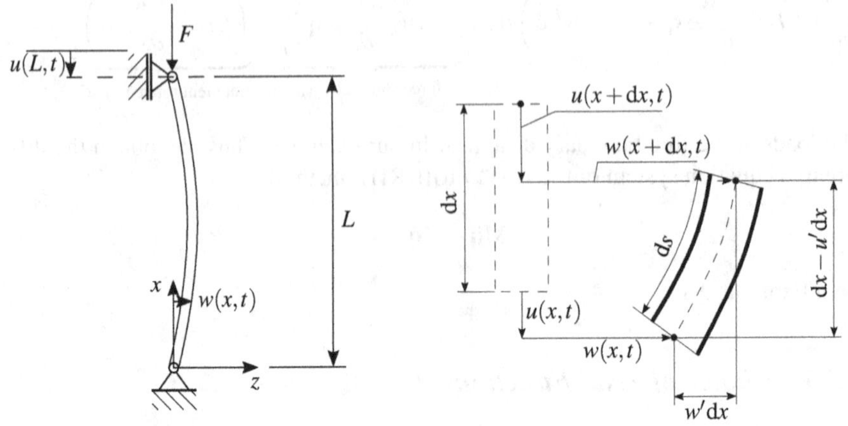

Fig. 3.14. Beam with longitudinal load and beam element.

3.3.6.1 Rayleigh-Ritz Method

Three energy terms are required for the RAYLEIGH-RITZ method:

- kinetic energy

$$T = \frac{1}{2}\rho A \int_0^L \dot{w}^2(x,t)\mathrm{d}x\,, \tag{3.124}$$

- bending potential

$$V = \frac{1}{2}EI \int_0^L w''^2(x,t)\mathrm{d}x\,, \tag{3.125}$$

- potential of the pressure force

$$V = -Fu(L,t) = -F\int_0^L \frac{1}{2}w'^2(x,t)\mathrm{d}x\,. \tag{3.126}$$

The potential of the pressure force decreases the stiffness and thus the bending potential for positive F. We express the longitudinal displacement $u(L)$ with the deflection w. Therefore, we need the following relations:

$$u(x+\mathrm{d}x,t) \approx u(x,t) + u'\mathrm{d}x\,, \tag{3.127}$$

$$w(x+\mathrm{d}x,t) \approx w(x,t) + w'\mathrm{d}x\,. \tag{3.128}$$

This yields (Fig. 3.14)

$$(\mathrm{d}x - u'\mathrm{d}x)^2 + (w'\mathrm{d}x)^2 = \mathrm{d}s^2 \tag{3.129}$$

neglecting the longitudinal deformation $\mathrm{d}s \approx \mathrm{d}x$

$$(1-u')^2 + w'^2 = 1\,. \tag{3.130}$$

We obtain a formula for

$$u' = 1 - \sqrt{1-w'^2} \approx \left(\frac{w'^2}{2}\right) \tag{3.131}$$

and finally

$$u(L,t) = \int_0^L u'\mathrm{d}x \approx \int_0^L \left(\frac{w'^2}{2}\right)\mathrm{d}x\,. \tag{3.132}$$

The approach

$$w(x,t) = \mathbf{q}(t)^T \mathbf{w}(x) \tag{3.133}$$

is inserted into the energy expressions:

$$T = \frac{1}{2}\dot{\mathbf{q}}^T \underbrace{\rho A \int_0^L \mathbf{w}(x)\mathbf{w}^T(x)\mathrm{d}x}_{\mathbf{M}}\, \dot{\mathbf{q}}\,, \tag{3.134}$$

$$V = \frac{1}{2}\mathbf{q}^T \underbrace{\left[EI\int_0^L \mathbf{w}''(x)\mathbf{w}''^T(x)\mathrm{d}x - F\int_0^L \mathbf{w}'\mathbf{w}'^T(x)\mathrm{d}x\right]}_{\mathbf{K}} \mathbf{q}\,. \tag{3.135}$$

With LAGRANGE's equations of the second kind, we obtain the equation of motion

$$\mathbf{M}\ddot{\mathbf{q}} + \mathbf{K}\mathbf{q} = \mathbf{0}\,. \tag{3.136}$$

The trial functions $\mathbf{w}(x)$ must satisfy the geometric boundary conditions:

$$\mathbf{w}(0) = \mathbf{0}\,, \quad \mathbf{w}(L) = \mathbf{0}\,. \tag{3.137}$$

If we select just one trial function, the matrices of the equation of motion become scalars. The trial function

$$w_1(x) = \sin\left(\frac{\pi x}{L}\right) \tag{3.138}$$

satisfies the required boundary conditions. With this trial function, we obtain:

$$M = \rho A \int_0^L \sin^2\left(\frac{\pi x}{L}\right) \mathrm{d}x = \frac{1}{2}\rho AL\,, \tag{3.139}$$

$$K = EI \int_0^L \left(\frac{\pi}{L}\right)^4 \sin^2\left(\frac{\pi x}{L}\right) \mathrm{d}x - F \int_0^L \left(\frac{\pi}{L}\right)^2 \cos^2\left(\frac{\pi x}{L}\right) \mathrm{d}x$$
$$= \frac{1}{2}EIL\left(\frac{\pi}{L}\right)^4 - \frac{1}{2}FL\left(\frac{\pi}{L}\right)^2 . \tag{3.140}$$

Thus, the eigenvalue problem $\det\left(M\lambda^2 + K\right) = 0$ transforms to

$$\frac{1}{2}\rho AL\lambda^2 + \left(\frac{\pi^2}{2L}\right)\left[EI\left(\frac{\pi}{L}\right)^2 - F\right] = 0\,. \tag{3.141}$$

With $\lambda^2 = -\omega^2$ for complex conjugate eigenvalues, we obtain the eigenfrequency, which depends on the pressure force F:

$$\omega = \sqrt{\left(\frac{\pi^2}{\rho AL^2}\right)\left[\frac{\pi^2 EI}{L^2} - F\right]}\,. \tag{3.142}$$

We study some cases below:

- The critical buckling load is reached when the eigen angular frequency disappears ($\omega = 0$). Then the re-acting force in the transverse direction vanishes:

$$F_{\mathrm{krit}} = \pi^2\left(\frac{EI}{L^2}\right)\,. \tag{3.143}$$

  Now, we can write

$$\omega^2 = \left(\frac{\pi^2}{\rho AL^2}\right)\left(F_{\mathrm{krit}} - F\right)\,. \tag{3.144}$$

- For $F = 0$, we obtain the eigen angular frequency of a beam which is simply supported on both sides:

$$\omega_0^2 = \frac{EI\pi^4}{\rho AL^4}\,. \tag{3.145}$$

- The eigen angular frequency of a prestressed string is obtained with vanishing bending stiffness $EI = 0$ and tensile forces $F < 0$:

$$\omega^2 = \frac{F\pi^2}{\rho AL^2}\,. \tag{3.146}$$

Fig. 3.15 shows the eigen angular frequency as a function of the pressure force F.

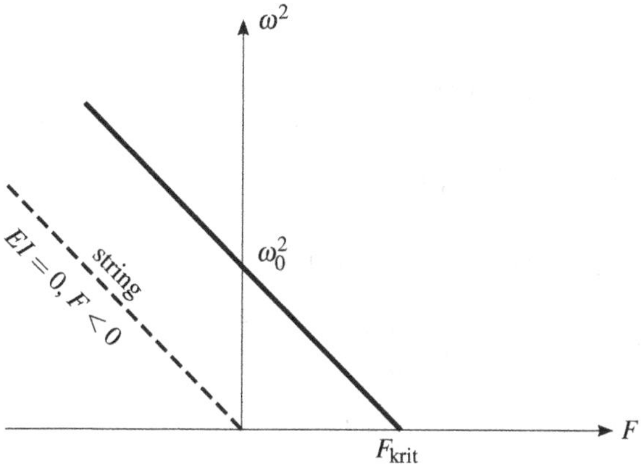

Fig. 3.15. Eigen angular frequency as a function of the pressure force F.

3.3.6.2 Bubnov-Galerkin Method

To apply the BUBNOV-GALERKIN method, we first derive the equations of motion using the beam element in Fig. 3.16. We need the following relations:

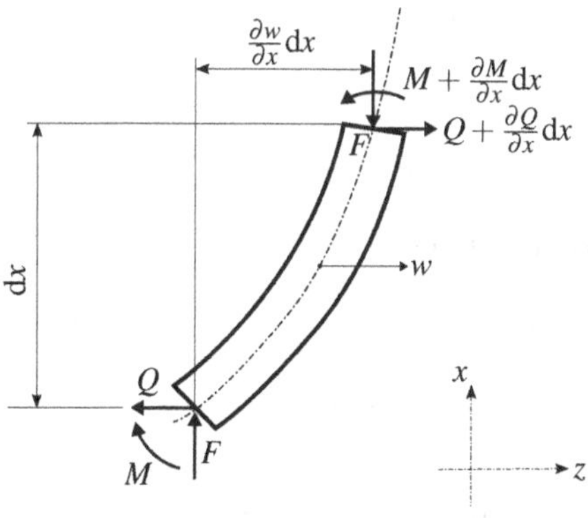

Fig. 3.16. Beam element.

- moment of momentum equation with neglected rotational inertia (about upper endpoint):

$$\left(\frac{\partial M}{\partial x}\right)\mathrm{d}x - Q\mathrm{d}x - F\left(\frac{\partial w}{\partial x}\right)\mathrm{d}x = 0\,, \tag{3.147}$$

- bending moment:

$$M = -EI\left(\frac{\partial^2 w}{\partial x^2}\right), \tag{3.148}$$

- momentum equation in the z-direction:

$$\rho A\left(\frac{\partial^2 w}{\partial t^2}\right)\mathrm{d}x = \left(\frac{\partial Q}{\partial x}\right)\mathrm{d}x\,. \tag{3.149}$$

Finally, we obtain the partial differential equation:

$$\rho A\frac{\partial^2 w(x,t)}{\partial t^2} + EI\frac{\partial^4 w(x,t)}{\partial x^4} + F\frac{\partial^2 w(x,t)}{\partial x^2} = 0\,. \tag{3.150}$$

Integration by parts in the method of weighted residuals

$$\int_0^L w_k\left(EIw^{(4)} + Fw'' + \rho A\ddot{w}\right)\mathrm{d}x = 0 \tag{3.151}$$

yields

$$\begin{aligned} 0 = &\int_0^L EIw_k''w''\mathrm{d}x + \underbrace{(EIw_k w''')_0^L}_{=0} - (w_k'M)_0^L \\ &- \int_0^L Fw_k'w'\mathrm{d}x + \underbrace{(Fw_k w')_0^L}_{=0} + \int_0^L \rho A w_k \ddot{w}\mathrm{d}x\,. \end{aligned} \tag{3.152}$$

The trial functions

$$w(x,t) = \mathbf{q}(t)^T\mathbf{w}(x) \tag{3.153}$$

must satisfy the geometric boundary conditions

$$\mathbf{w}(0) = \mathbf{0}\,, \quad \mathbf{w}(L) = \mathbf{0}\,. \tag{3.154}$$

Inserting the free boundary conditions $(M)_0^L = 0$ yields

$$\underbrace{\rho A\int_0^L \mathbf{w}\mathbf{w}^T\mathrm{d}x}_{\mathbf{M}}\ddot{\mathbf{q}} + \underbrace{\left[EI\int_0^L \mathbf{w}''\mathbf{w}''^T\mathrm{d}x - F\int_0^L \mathbf{w}'\mathbf{w}'^T\mathrm{d}x\right]}_{\mathbf{K}}\mathbf{q} = \mathbf{0}\,. \tag{3.155}$$

This gives the same equation of motion as for the RAYLEIGH-RITZ method:

$$\mathbf{M}\ddot{\mathbf{q}} + \mathbf{K}\mathbf{q} = \mathbf{0}\,. \tag{3.156}$$

3.4 Vibrations of Elastic Multibody Systems

Most mechanical systems in practice consist of both, bodies, which we can assume to be rigid, and bodies, the deformation of which must be considered. From the many techniques for the representation of such systems, we choose, seen from an engineering point of view, the one that offers maximum transparency and closeness to reality with minimum effort. A measure is the minimum number of degrees of freedom, which just describe the motion of the system adequately.

The division into rigid and nonrigid bodies already guarantees to minimize the degrees of freedom in a first step, because by modeling a part of the system components as rigid bodies the smallest possible number of degrees of freedom is used. In the second step, the nonrigid system components have to be modeled with a minimum number of degrees of freedom such that satisfactory realism can be achieved. For linear elastic systems, the presented methods of RAYLEIGH-RITZ and BUBNOV-GALERKIN are based on global (modal) trial functions, because only a small number of trial functions and thus additional elastic degrees of freedom are necessary due to always existing structural damping and its increasing effect on higher eigenmodes (dissipation energy). One important criterion for selecting the elastic modes are the frequencies of operation of a machine or a structure. Only the elasticity of components is of interest, the eigenfrequencies of which are in the range of these operating frequencies.

The equations of motion of these systems can be derived with the methods presented in Chapter 1. The equations for rigid multibody systems must be extended by the nonrigid components. For the deformations of linear-elastic systems an approximate approach in the sense of RAYLEIGH-RITZ or BUBNOV-GALERKIN can always be found, which accurately describes the individual deformations. Without going into the mathematical details, the procedure will be outlined below:

1. Setting up a mechanical model.
2. Defining the rigid and elastic (nonrigid) bodies.
3. Finding the best coordinate systems.
4. Setting up the free-body diagram with relevant forces.
5. Evaluating the relative kinematics with all deformation effects (absolute velocities and accelerations, transformation matrices).
6. Discretization of the elastic components (choice of trial functions, number of elastic degrees of freedom).
7. Choice of generalized coordinates for the rigid bodies.
8. Determination of JACOBIANs of translation and rotation.
9. Derivation of the projected equations of motion.
10. Analytical and numerical treatment of the equations of motion (first integrals, linearization, analytical or numerical solution).

A detailed consideration of such problems can be found in [11, 10, 59, 51].

Fig. 3.17 shows a typical example of an elastic multibody system [55]. The Ravigneaux planetary gearset is used in automatic transmissions of motor vehicles

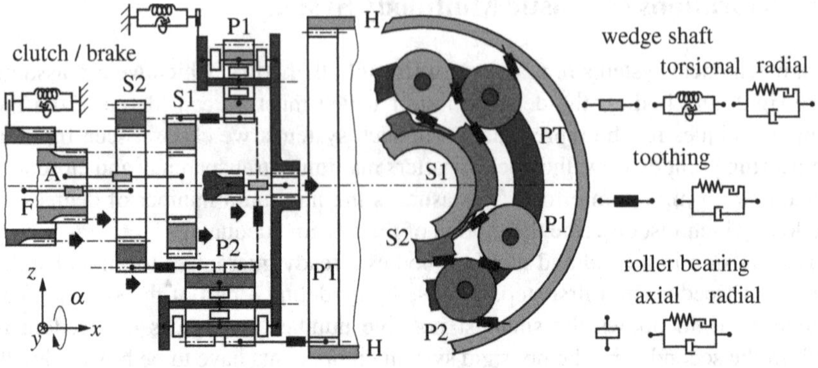

Fig. 3.17. Mechanical model of a Ravigneaux planetary gearset [55].

primarily because of the small installation space and the variety of switching options. By selecting different drive shafts, different overall transmission ratios can be realized.

The main components of the Ravigneaux planetary gearset are a small sun $S1$ and a large sun $S2$ (small central gears), a ring H (large central gear), a planet carrier PT (rack) with short planets $P1$ and long planets $P2$, the drive shafts A, and the free coupling shafts F. The large sun $S2$, the planets $P2$, and the ring H together with the planet carrier PT represent a simple planetary gear. The inner planets $P1$ are mounted on the same rack. The power is transmitted through the long planets $P2$ to the common ring H.

In the planetary gear shown in Fig. 3.17, the output goes via the ring H. For the coupling of the ring H to the output shaft, several design variations exist. Very often, the ring H is welded to the output shaft. In this case, the planar elastic deformation of the ring is very small due to the radial stiffening; it is therefore allowed to model the ring as a rigid body. However, if a thin-walled ring is coupled to the output via a gear teeth set according to Fig. 3.17, it may deform due to the presence of radial and flank clearance and must be modelled as an elastic body. Analogously to simple planetary gearsets, the gears and the rack PT are modelled as rigid bodies with 4 or 6 degrees of freedom. For the reduced planetary gearset, the coupling of the individual bodies with each other and with the input and output is modelled by tooth couplings, bearings, and wedge shaft connections.

Chapter 4
Methods for Nonlinear Mechanics

4.1 General Remarks

The motion of any mechanical system composed of rigid bodies or discretized elastic bodies is always described by a set of nonlinear ordinary differential equations of second order of the form (Chapter 1)

$$\mathbf{M}(\mathbf{q},t)\ddot{\mathbf{q}} = \mathbf{h}(\mathbf{q},\dot{\mathbf{q}},t) \ . \tag{4.1}$$

The vector $\mathbf{h}$ contains force or torque expressions, which might depend on velocities $\dot{\mathbf{q}}$ of maximum second order. In spite of the simplicity of these equations, it is impossible to find a general solution or to find some *superposition principle*, which allows us to combine some fundamental solutions like in the linear case (Chapter 2). Though, such a general solution can be evaluated for one-dimensional systems, for which the equation of motion is of RICCATI type [33]. For systems with many degrees of freedom and also for continuum mechanical problems, one cannot avoid numerical integration routines and/or approximations.

With respect to approximations, one makes use of special features of the differential equations or of the mechanical system under consideration. This concerns systems with *weak nonlinearities* or *periodical systems*. Considering the first case, we may develop the equations of motion with respect to a small parameter resulting in a sequence of differential equations, where each new set represents an improved description of the problem compared with the preceding set, convergence anticipated [47, 48, 26].

Considering nonlinear vibration systems, we make use of their periodicity

$$\mathbf{q}(t) = \mathbf{q}(t+T), \tag{4.2}$$

where the period T usually is given by nonlinear equations. A couple of methods are available for their solution, particularly in connection with limit cycles. We discuss some simple examples.

We distinguish *quantitative* and *qualitative* methods. Qualitative methods establish some statements concerning general solution properties by geometric

F. Pfeiffer and T. Schindler, *Introduction to Dynamics*,
DOI: 10.1007/978-3-662-46721-3_4

considerations of the equations of motion. Some aspects are stability of motion, periodical behavior, *bifurcations* of the solution, and possible chaotic developments of the motion. They are connected with names like POINCARÉ, LYAPUNOV, THOM, THOMPSON, ARNOLD, and others. Also in the future, geometric considerations of the nonlinear equations of motion will contribute significantly to the understanding of the phenomena included by them and from this also to many important applications of practical relevancy [71, 3, 70, 75].

Such a textbook cannot treat nonlinear dynamics in an exhaustive way. We recommend special literature [71, 3, 24, 26, 33, 39, 44]. However with respect to a first introduction, we consider some important aspects, which might help when looking at practical problems. In a first step, we apply four methods to a simple nonlinear oscillator with one degree of freedom (DOF) only. These four methods are

- piecewise exact solutions,
- weighted residuals,
- harmonic balance,
- least squares.

As an example of the qualitative methods, we consider LYAPUNOV's stability theory.

With respect to practice, analytical solutions and approximations make sense only if they help to understand the problems under consideration by a deeper insight into their structures and their mechanical features. This requires a careful choice of methods, especially in the case of nonlinear systems. Not always, it will be advisable to use numerical methods leading sometimes to problems of interpretation. Very simple models often generate information that is hidden in the results of large models requiring long investigations. This situation justifies our presentation of simple methods applied to a simple example with one degree of freedom.

4.2 Phase Space

For a better understanding of dynamics we often use *phase portraits* drawn in a *phase plane*. For this purpose, we select from the *state space* two state trajectories over time and sketch them with respect to each other (projected phase curve). Usually, we select velocity and position states. This type of presentation is also well suited to nonsmooth systems. For a one-DOF example with the equation of motion

$$\ddot{x} = -\omega^2 x\,, \tag{4.3}$$

the initial position x_0 and vanishing initial velocity, we discuss three methods for achieving a *phase portrait*. In this case, this corresponds to the whole set of phase curves.

1. *Solution of the equations of motion and elimination of time*

$$x(t) = +x_0\cos(\omega t)\,, \tag{4.4}$$

$$\dot{x}(t) = -x_0\omega\sin(\omega t) \tag{4.5}$$

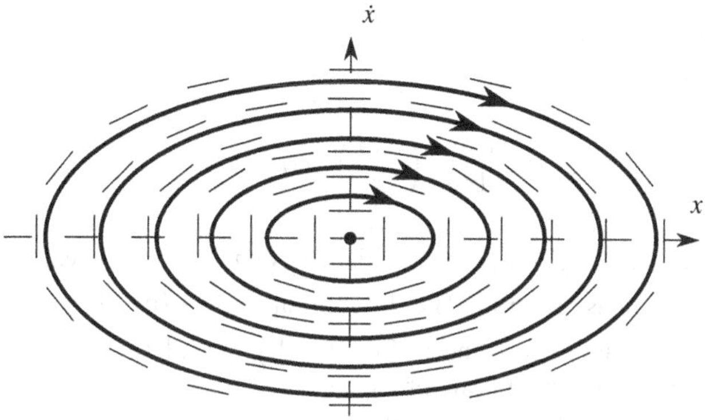

Fig. 4.1. Phase portrait of a 1-DOF oscillator.

results in

$$\left(\frac{x}{x_0}\right)^2 + \left(\frac{\dot{x}}{\omega x_0}\right)^2 = \cos^2(\omega t) + \sin^2(\omega t) = 1\,. \tag{4.6}$$

These equations describe ellipses in the phase plane (Fig. 4.1). For the *stationary point* $\begin{pmatrix} x & \dot{x} \end{pmatrix}^T = \begin{pmatrix} 0 & 0 \end{pmatrix}^T$, the corresponding trajectory remains there for all time. The sense of rotation of these trajectories follows from the functional relation of x and $\dot{x}$: for example, the value of x increases for a positive velocity $\dot{x} > 0$. The phase curves will not have an intersection as long as the right-hand side of the initial value problem is smooth enough. Every phase point in Fig. 4.1 is part of one phase curve only. Therefore, Fig. 4.1 completely describes the phase portrait. However, we should keep in mind that phase portraits represent a plane cut of the total phase space. In general, intersecting projected phase curves are possible (see for example the book cover).

2. *Integration of the slope and conservation of energy*
   The slope of a phase curve

$$\frac{\mathrm{d}\dot{x}}{\mathrm{d}x} = \frac{\ddot{x}}{\dot{x}} = -\frac{\omega^2 x}{\dot{x}} \tag{4.7}$$

   can be remodeled by separation of variables:

$$\omega^2 x \mathrm{d}x = -\dot{x}\mathrm{d}\dot{x}\,. \tag{4.8}$$

   Integration results in

$$\frac{1}{2}\omega^2 x^2 + \frac{1}{2}\dot{x}^2 = \frac{1}{2}\omega^2 x_0^2 = \frac{E}{m}\,, \tag{4.9}$$

   which is the already considered ellipse equation. Every ellipse in the phase space indicates conservation of energy.

3. *Isoclines (phase velocity field)*
   According to (4.7), it is

$$\begin{pmatrix} \mathrm{d}x \\ \mathrm{d}\dot{x} \end{pmatrix}^T \begin{pmatrix} \omega^2 x \\ \dot{x} \end{pmatrix} = 0\,, \tag{4.10}$$

   which says, that $\left(\mathrm{d}x \;\; \mathrm{d}\dot{x}\right)^T$ is perpendicular to $\left(\omega^2 x \;\; \dot{x}\right)^T$. This feature allows one to construct a phase portrait. The differential change $\left(\mathrm{d}x \;\; \mathrm{d}\dot{x}\right)^T$ for a point $\left(x \;\; \dot{x}\right)^T$ is indicated as a short line in Fig. 4.1. By considering all these small lines, we also come out with elliptic phase curves.
   For an intersection point $\left(x \;\; \dot{x}\right)^T = \left(x \;\; 0\right)^T$, we get

$$\mathrm{d}x = -\frac{\dot{x}}{\omega^2 x}\mathrm{d}\dot{x} = 0\,, \tag{4.11}$$

   which means that phase curves intersect the x-axis always in a perpendicular way.

4.3 A 1-DOF Nonlinear Oscillator

We consider a 1-DOF oscillator with a nonlinear restoring force $r(x)$ (Fig. 4.2). No other forces are applied, and the equations of motion write

$$m\ddot{x} = -r(x)\,. \tag{4.12}$$

With $\dot{x} = \frac{\mathrm{d}x}{\mathrm{d}t}$ (*definition velocity*) and $\mathrm{d}t = \frac{1}{\dot{x}}\mathrm{d}x$ (*elimination of time*), we get

$$m\dot{x}\mathrm{d}\dot{x} = -r(x)\mathrm{d}x\,. \tag{4.13}$$

After the *separation* of x and $\dot{x}$, an integration is possible and results in

$$\frac{1}{2}m\dot{x}^2 = E_0 - \int r(x)\mathrm{d}x \tag{4.14}$$

with $E_0 := \frac{1}{2}m\dot{x}_0^2$. From this

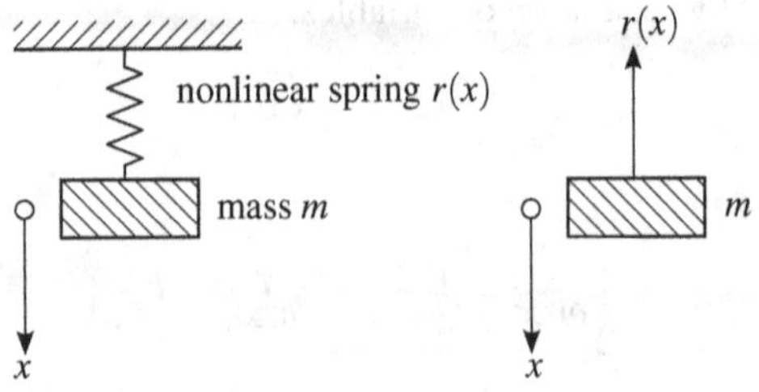

Fig. 4.2. A 1-DOF nonlinear oscillator.

$$\dot{x} = \sqrt{\frac{2}{m}\left(E_0 - \int r(x)\mathrm{d}x\right)} \tag{4.15}$$

and finally

$$t(x) = t_0 + \int_0^x \frac{\mathrm{d}x}{\dot{x}} , \tag{4.16}$$

we determine the reversal $x(t)$.

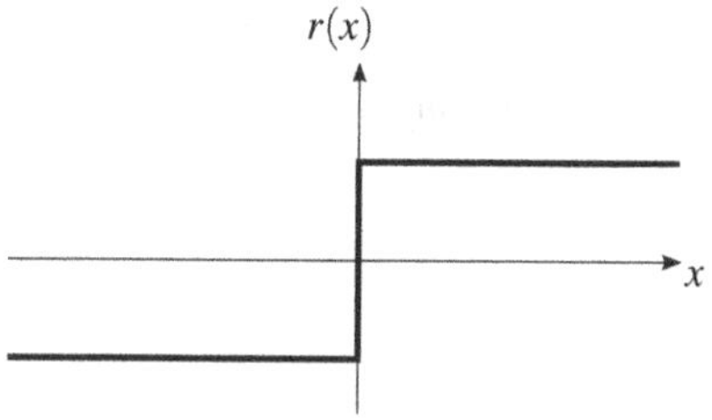

Fig. 4.3. Signum function.

These relations are general and will be used for the four methods discussed in the following. For this purpose, we choose a specific nonlinear restoring force

$$r(x) = r \cdot \mathrm{sgn}(x) , \tag{4.17}$$

with constant r. It is depicted in Fig. 4.3.

Example 4.1 (Piecewise constant restoring force and corresponding solution). The following three nonlinear oscillators possess restoring forces according to (4.17). It is very easy to test that by simple experiments. The first example of a rod rocking on a block from left to right and from right to left can easily be calculated, if we neglect the impact losses accompanying this oscillation (Fig. 4.5). The moment of momentum equation about R is

$$J_R \ddot{\alpha} = -\frac{mgb}{2} . \tag{4.18}$$

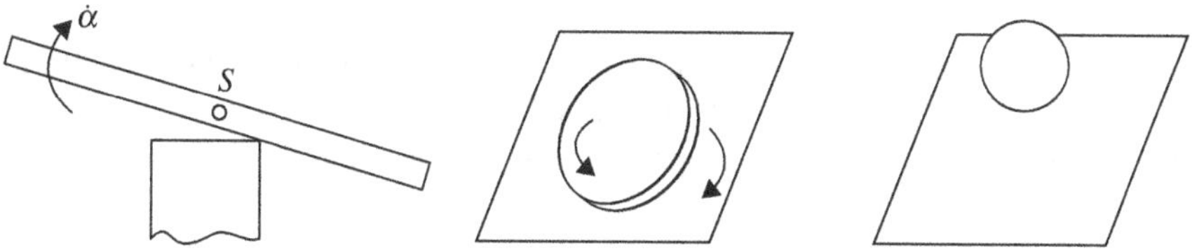

Fig. 4.4. From the left: rod on block, coin on plane, ball on plane.

If the rod is rocking on point L we have, using a relevant definition of the angle α,

$$J_L\ddot{\alpha} = \frac{mgb}{2}\,. \tag{4.19}$$

The moments of inertia are

$$J = J_R = J_L = J_S + m\left(\frac{b}{2}\right)^2 = \frac{ml^2}{12}\left(1+3\left(\frac{b}{l}\right)^2\right)\,, \tag{4.20}$$

and combining these relations, we get the 1-DOF oscillator with a nonlinear restoring force according to (4.17)

$$J\ddot{\alpha} = -\left(\frac{1}{2}mgb\right)\operatorname{sgn}(\alpha)\,. \tag{4.21}$$

First, we consider (4.18):

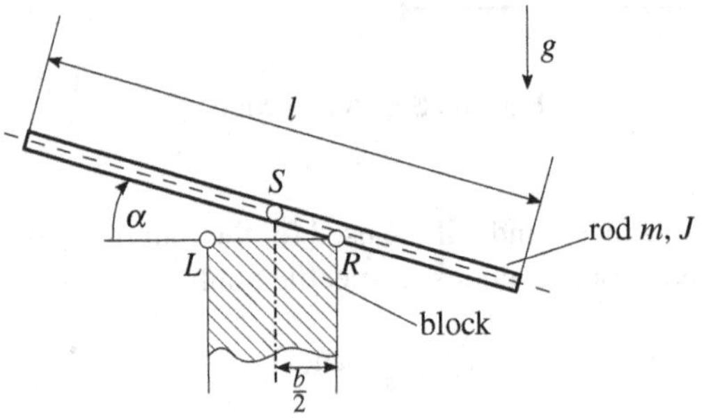

Fig. 4.5. Rod on block.

$$\ddot{\alpha} = -\frac{r}{J} = -\frac{6bg}{l^2}\left(\frac{1}{1+3\left(\frac{b}{l}\right)^2}\right)\,. \tag{4.22}$$

Formal integration results in

$$\dot{\alpha} = -\frac{r}{J}(t-t_0) + C_1\,, \tag{4.23}$$

$$\alpha = -\frac{1}{2}\frac{r}{J}(t-t_0)^2 + C_1(t-t_0) + C_2\,. \tag{4.24}$$

The constant magnitudes C_1 and C_2 are determined by the initial conditions at time $t = t_0$, at $\dot{\alpha}|_{t_0} = \dot{\alpha}_0$, and at $\alpha|_{t_0} = 0$. They give $C_1 = \dot{\alpha}_0$ and $C_2 = 0$. Introducing the velocity, eliminating the time, and separating the variables result in $\dot{\alpha}\frac{d\dot{\alpha}}{d\alpha} = \ddot{\alpha} = -\frac{r}{J}$ and from this

$$\dot{\alpha}^2 = -2\frac{r}{J}\alpha + \alpha_0^2\,. \tag{4.25}$$

Equation (4.25) represents a parabola

$$\dot{\alpha} = \sqrt{\dot{\alpha}_0^2 - 2(\frac{r}{J})\alpha} \tag{4.26}$$

in the phase plane $(\alpha, \dot{\alpha})$, the second half of which can be added by the corresponding solution of (4.19) (Section 4.3.1 and Fig. 4.6).

4.3.1 *Piecewise Exact Solution*

With respect to the piecewise constant feedback function $r(x) = r\text{sgn}(x)$, there exist only two areas of motion, each of which is characterized by equations of motion with constant coefficients. They are

$$m\ddot{x} = -r \text{ if } x > 0\,, \tag{4.27}$$

$$m\ddot{x} = +r \text{ if } x < 0\,. \tag{4.28}$$

For these simple equations an analytical solution by elementary integration is feasible, and the two solutions must then be put together. For the case $x > 0$, we get

$$\ddot{x} = -(\frac{r}{m})\,, \tag{4.29}$$

$$\dot{x} = -(\frac{r}{m})t + \dot{x}_0\,, \tag{4.30}$$

$$x = -\frac{1}{2}(\frac{r}{m})t^2 + \dot{x}_0 t + x_0\,. \tag{4.31}$$

Elimination of time from the velocity solution and regarding $x_0 = 0$ and $t_0 = 0$ result in

$$x = (\frac{m}{2r})\,(\dot{x}_0^2 - \dot{x}^2)\,, \tag{4.32}$$

which defines a parabola in the phase plane $(x, \dot{x})$ (Fig. 4.6). The transient response described by (4.31) represents a parabola, too, with the characteristic magnitudes (Fig. 4.6)

$$t_1 = \frac{2\dot{x}_0 m}{r}\,, \quad t_m = \frac{t_1}{2} = \frac{\dot{x}_0 m}{r}\,, \quad A = x(t_m) = \frac{\dot{x}_0^2 m}{2r}\,. \tag{4.33}$$

At point $t = t_1$, the sign of x becomes negative ($\text{sgn}(x) < 0$), and the *exact solution* continues by the same parabola but mirrored. From this result, the period satisfies

$$T = 2t_1 = \frac{4\dot{x}_0 m}{r}\,. \tag{4.34}$$

Expressing the initial velocity $\dot{x}_0$ by the amplitude A

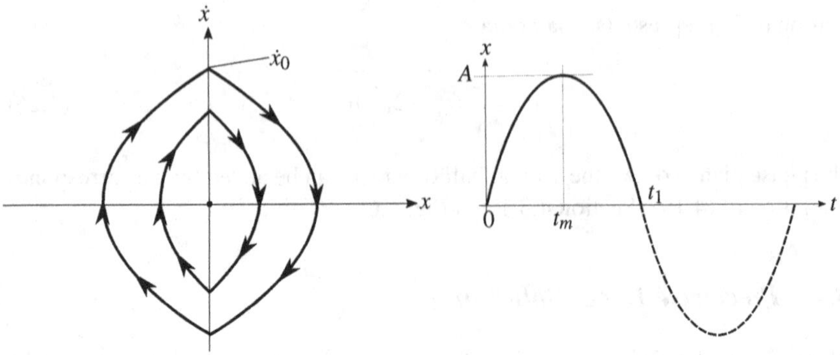

Fig. 4.6. Phase plane $(x,\dot{x})$ and time behavior $x(t)$.

$$\dot{x}_0 = \sqrt{\frac{2Ar}{m}} \tag{4.35}$$

comes out with

$$T = \frac{4m}{r}\sqrt{\frac{2Ar}{m}} = \sqrt{\frac{32mA}{r}} = 5,66\sqrt{\frac{mA}{r}}\,. \tag{4.36}$$

The period T depends on the amplitude A, which is a characteristic feature of nonlinear oscillations (Fig. 4.7). An experiment with a thrown coin will confirm that immediately (Fig. 4.4). The dependence of the period on the amplitude may be an important indication for the assessment of vibrations.

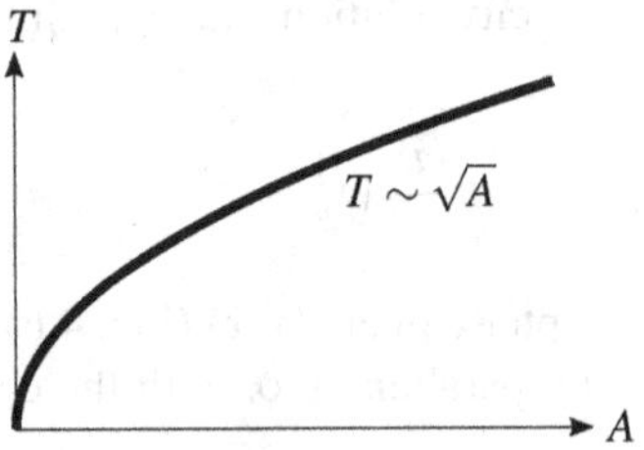

Fig. 4.7. Characteristic relationship $T(A)$.

4.3.2 *Method of Weighted Residuals*

The basic idea of this method has already been presented in Section 2.3 considering examples of linear systems. In the following, we demonstrate its applicability with respect to nonlinear systems. For this purpose, we approximate the solution $x(t)$ of a time-dependent nonlinear differential equation

$$D[x] = 0 \tag{4.37}$$

by the ansatz

$$x_N(t) = \mathbf{a}^T \mathbf{w}(t) \tag{4.38}$$

with the constant coefficients a_i and the trial functions $w_i(t)$. The unknown coefficients a_i are evaluated in such a way, that the weighted residual

$$\int w_k(t) D\left[\mathbf{a}^T \mathbf{w}(t)\right] \mathrm{d}t = 0 \tag{4.39}$$

becomes zero. The method is very general and can be used for time-dependent as well as for space-dependent problems.

We again consider the one-dimensional example $m\ddot{x} = -r(x)$ with a piecewise constant function $r(x) = r\mathrm{sgn}(x)$. Applying the ansatz

$$x(t) = A\sin(\omega t) \tag{4.40}$$

and applying the method of the weighted residuals, we arrive at

$$\int_0^{\frac{2\pi}{\omega}} [m\ddot{x} + r\mathrm{sgn}(x)] \sin(\omega t)\, \mathrm{d}t = 0\,. \tag{4.41}$$

Considering the periodicity of the problem, we integrate over one cycle $T = \frac{2\pi}{\omega}$ only. Now, we separate the integral into the two parts of the signum function resulting in (Fig. 4.8)

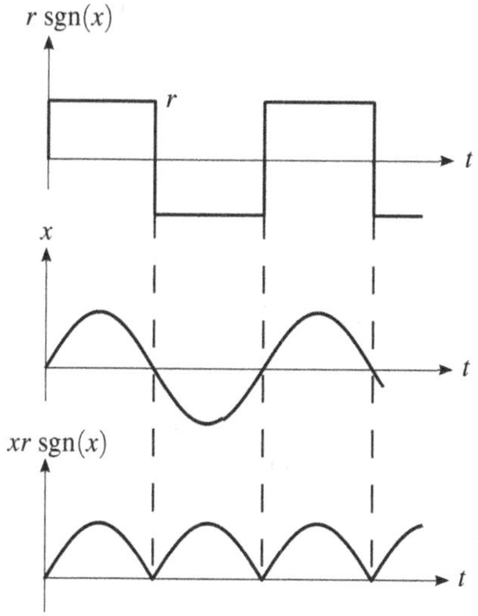

Fig. 4.8. Properties of the signum function.

$$\int_0^{\frac{\pi}{\omega}} \left[-mA\omega^2 \sin(\omega t) + r\right] \sin(\omega t)\,\mathrm{d}t + \int_{\frac{\pi}{\omega}}^{\frac{2\pi}{\omega}} \left[-mA\omega^2 \sin(\omega t) - r\right] \sin(\omega t)\,\mathrm{d}t = 0 \tag{4.42}$$

and after performing the integration

$$-2\left[mA\omega^2\left(\frac{\pi}{2\omega}\right)\right] + 2\left[\frac{2r}{\omega}\right] = 0\,. \tag{4.43}$$

From this, we get the eigenfrequency as

$$\omega^2 = \frac{4r}{\pi mA}\,, \tag{4.44}$$

$$T = \frac{2\pi}{\omega} = \sqrt{\frac{\pi^3 mA}{r}} = 5{,}57\sqrt{\frac{mA}{r}}\,. \tag{4.45}$$

It results in an error in comparison with $T = 5{,}66\sqrt{\frac{mA}{r}}$ in (4.36) of the piecewise exact solution of only $1{,}6\%$.

4.3.3 *Harmonic Balance*

What we call *harmonic balance* is sometimes also described by *harmonic linearization*. This approach replaces the original nonlinear differential equation by a linear differential equation, which approximates in the best way the motion over one period. Starting again with the standard example of the previous sections, this means to replace the equation

$$m\ddot{x} + r\mathrm{sgn}(x) = 0 \tag{4.46}$$

by the approximation

$$m\ddot{x} + cx = 0, \tag{4.47}$$

where in this case the term cx stands for the piecewise constant function $r(x)$. It is an odd function.

$$r(x) = -r(-x)\,, \tag{4.48}$$

$$r(0) = 0\,. \tag{4.49}$$

Therefore, we also use an odd ansatz

$$x(t) = A\sin(\omega t) \tag{4.50}$$

and expand a FOURIER series for the odd function $r(x)$

$$r(x) = \sum_{\nu=1}^{\infty} a_\nu \sin(\nu\omega t) \; . \tag{4.51}$$

Considering only the fundamental harmonic

$$cA\sin(\omega t) = cx = r(x) \approx a_1 \sin(\omega t) \; , \tag{4.52}$$

we get a relation for the *spare coefficient c*

$$c = \frac{a_1}{A} \; . \tag{4.53}$$

As $r(x)$ is an odd function, the spare coefficient follows from

$$a_1 = 2(\frac{2}{T}) \int_0^{\frac{T}{2}} r\mathrm{sgn}(A\sin(\omega t))\sin(\omega t)\,\mathrm{d}t \; . \tag{4.54}$$

In the interval under consideration (Fig. 4.8), the sign of $A\sin(\omega t)$ is positive,

$$a_1 = \frac{4r}{T} \int_0^{\frac{T}{2}} \sin(\omega t)\,\mathrm{d}t = -\frac{2r}{\pi}\cos(\frac{2\pi}{T})\bigg|_0^{\frac{T}{2}} = \frac{4r}{\pi} \; , \tag{4.55}$$

and from this, we come to

$$c = \frac{a_1}{A} = \frac{4r}{\pi A} \tag{4.56}$$

and to the linear differential equation approximating the original one

$$m\ddot{x} + \left(\frac{4r}{\pi A}\right) x = 0 \; . \tag{4.57}$$

The eigenfrequency is

$$\omega^2 = \frac{4r}{\pi m A} \; , \tag{4.58}$$

$$T = \frac{2\pi}{\omega} = \sqrt{\frac{\pi^3 m A}{r}} = 5{,}57\sqrt{\frac{mA}{r}} \; . \tag{4.59}$$

It is the same result as in the case of the weighted residuals, originating from the fact, that behind all these approximations there is a more or less hidden least square idea.

4.3.4 Method of Least Squares

As before we want to generate a linear spare differential equation, but now by minimizing the difference between the original and the spare differential equation

$$\Delta = [m\ddot{x} + r(x)] - [m\ddot{x} + cx] = r(x) - cx \tag{4.60}$$

in the sense of least squares integrated over one period

$$\bar{\Delta}^2 = \frac{1}{T}\int_0^T [r(x) - cx]^2 \,\mathrm{d}t \to \min\,. \tag{4.61}$$

A necessary condition for a minimum is

$$0 = \left(\frac{\partial \bar{\Delta}^2}{\partial c}\right) = \frac{2}{T}\int_0^T [r(x) - cx]\,x\mathrm{d}t\,, \tag{4.62}$$

which results in

$$c = \frac{\int_0^T r(x)x\mathrm{d}t}{\int_0^T x^2\mathrm{d}t}\,. \tag{4.63}$$

As above, we choose the ansatz $x(t)$

$$x(t) = A\sin(\omega t) \tag{4.64}$$

and insert it into (4.63)

$$c = \frac{\int_0^{\frac{T}{2}} rA\sin(\omega t)\,\mathrm{d}t - \int_{\frac{T}{2}}^{T} rA\sin(\omega t)\,\mathrm{d}t}{\int_0^T A^2\sin^2(\omega t)\,\mathrm{d}t} = \frac{4r}{\pi A}\,, \tag{4.65}$$

which comes out with the same results as before, because of the arguments given above.

4.3.5 *Practical Example*

A highly illustrative example of some of the remarkable properties of nonlinear oscillations generated by gear rattling has been detected in an application connected with noise problems of a pleasure boat [52]. The boat is equipped with a drive train system combining a 340-hp eight-cylinder engine with a reversal gear, which in one direction operates as a one-stage gear system and in the other direction as a two-stage gear system (Fig. 4.9). To switch between rotation and counter-rotation, a clutch is used which possesses backlash. In addition, all gear meshes also have backlashes. Therefore, and dependent on the switched state, the system can rattle, the rattling process being excited by large engine unbalances. The rattling was occasionally so loud that the drive system could not be sold. Several parameter studies were performed applying classical theories, but parameter variations only marginally improved the noise behavior. A breakthrough occurred as a result of an idea of the company's design engineering department, which used a special clutch with a maximum angular backlash of up to 35°. The experiments indicated a noise

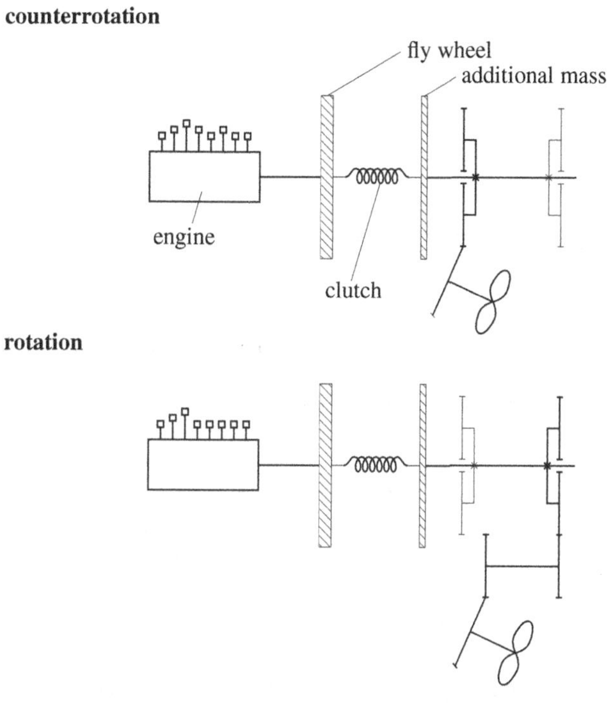

Fig. 4.9. Scheme of the ship turning gear.

breakdown for an angular play of about 17°. Simulations have confirmed this magnitude. Figure 4.10a displays the principal situation.

One can explain this strange behavior by considering a very simple 1-DOF model with backlash. The system is periodically excited and has a nonlinearity due to a spring with backlash that corresponds to the mesh of the gear teeth (Fig. 4.10b).

The equation of motion in dimensionless form is

$$\xi'' + D\xi' + \varphi(\xi) = \varphi_0 \cos\tau \tag{4.66}$$

with the magnitudes $\xi = \frac{x}{v}$, $\tau = \omega t$, $(\cdot)' = \frac{\mathrm{d}}{\mathrm{d}\tau}$, $D = \frac{d}{m\omega}$, $\varphi(\xi) = \frac{F(x)}{m\omega^2 v}$, $\varphi_0 = \frac{F_0}{m\omega^2 v}$, and $\eta = \frac{c}{m\omega^2}$. The nonlinear force law $F(x)$ is then

$$\varphi(\xi) = \eta(\xi - \frac{1}{2}) \quad \text{for} \quad \xi > +\frac{1}{2}, \tag{4.67}$$

$$\varphi(\xi) = \eta(\xi + \frac{1}{2}) \quad \text{for} \quad \xi < -\frac{1}{2}, \tag{4.68}$$

$$\varphi(\xi) = 0 \qquad \text{for} \quad -\frac{1}{2} \leq \xi \leq +\frac{1}{2}. \tag{4.69}$$

We approximate the nonlinear equation (4.66) by the linear one

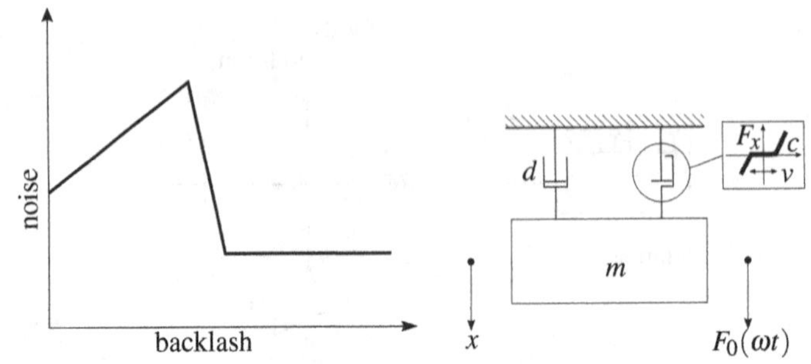

(a) Noise over backlash. (b) 1-DOF oscillator with backlash.

Fig. 4.10. Analysis of the ship turning gear.

$$\xi'' + D\xi' + \eta_0 \xi = \varphi_0 \cos\tau \tag{4.70}$$

and evaluate η_0 by a least square method, taking into account (4.67)-(4.69). This results in

$$\eta_0 = \left(\frac{2\eta}{\pi\xi_0}\right)\left[\xi_0 \arccos\left(\frac{1}{2\xi_0}\right) - \frac{1}{2}\sqrt{1-\left(\frac{1}{2\xi_0}\right)^2}\,\right], \tag{4.71}$$

where $\xi_0 = \frac{x_0}{v}$ is the gain of the solution of (4.70).

Gain and phase ψ are

$$\frac{\xi_0}{\varphi_0} = m\omega^2 \frac{x_0}{F_0} = \frac{1}{\sqrt{(1-\eta_0)^2 + D^2}}, \tag{4.72}$$

$$\tan\psi = \frac{D}{1-\eta_0}. \tag{4.73}$$

Figure 4.11 depicts the resonance curves for this approximation. They reveal some astonishing properties. The diagram on the left of Fig. 4.11 illustrates the well-known behavior of resonance curves for systems with backlash. Scanning the play characteristic (4.67)-(4.69) from zero force to nonzero force results in a resonance structure that is "more than linear" (resonance peak turns to the right); passing from the nonzero force branch to zero force we obtain a structure that is "less than linear" (resonance peak turns to the left). Both effects can be seen in Fig. 4.11. With increasing dimensionless play $\frac{v}{F_0/c}$, the jump phenomenon is significantly intensified when increasing or decreasing the excitation frequency $\frac{\Omega}{\sqrt{c/m}}$.

The diagram on the right of Fig. 4.11 indicates a very strong jump behavior depending on the backlash itself. If the backlash v is of the order of magnitude of the spring deflection $\frac{F_0}{c}$ caused by the excitation force amplitude F_0, we get a steep descent of the amplification factor $\frac{x_0}{F_0/c}$, where the character of this descent depends

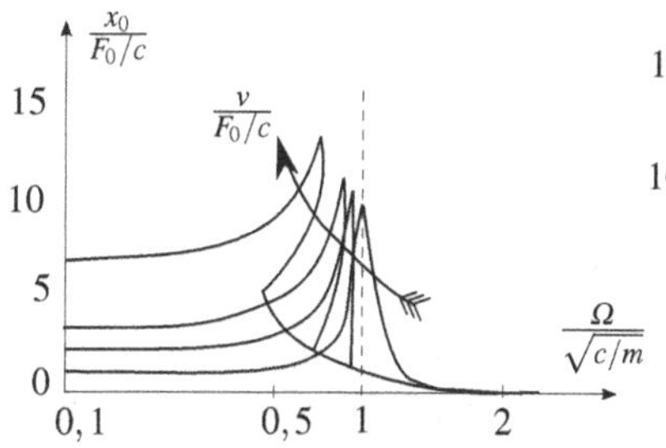

(a) Nonlinear resonance curves.

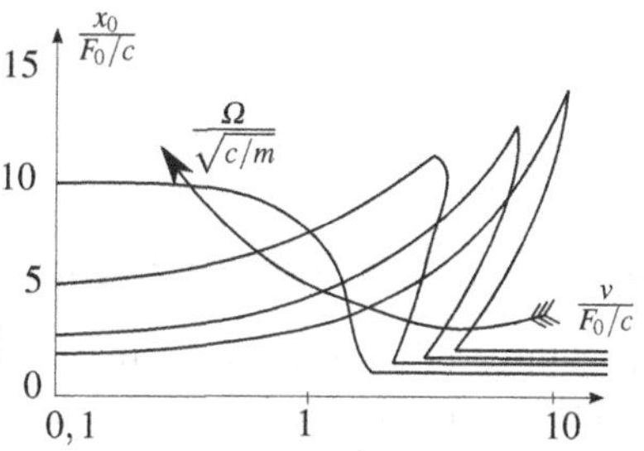

(b) Jump behavior.

Fig. 4.11. Important parameter influences concerning rattling of a ship turning gear.

on the excitation frequency. With decreasing $\frac{\Omega}{\sqrt{c/m}}$, we see a significant shift of the amplification peaks to the right. The amplification $\frac{x_0}{F_0/c}$ is proportional to the noise amplitudes. Therefore, the results of Fig. 4.11 give a physical interpretation of the rattling phenomena as measured in the ship reversal gear. In addition, this could be confirmed by a simulation with a complete model, which indicates that the neglections in the simplified model are correctly estimated.

4.4 Stability of Motion

A confusingly large number of *stability terms* exist, terms like linear, nonlinear, static, dynamic, global, total, weak, strong, or asymptotic stability [38, 27, 33, 45]. In the following, we focus our attention on

- stability of states of rest,
- stability of orbits.

The concept of some *norm* plays an important role with respect to general stability definitions. For deviations from a reference state, however it may be defined, we need some non-negative measure. The analysis of the bounds of such measures allows some insight into the global behavior of motion of a dynamic system and thus some statements with respect to its stability of motion. Some examples of norms are the Euclidean norm , the weighted Euclidean norm and the arithmetic norm :

$$\| \mathbf{x} \|_2 = \sqrt{\sum x_i^2} = \sqrt{\mathbf{x}^T \mathbf{x}}\,, \tag{4.74}$$

$$\| \mathbf{x} \|_{2,\mathbf{R}} = \sqrt{\mathbf{x}^T \mathbf{R} \mathbf{x}} \quad \text{with} \quad \mathbf{R} = \mathbf{R}^T > 0\,, \tag{4.75}$$

$$\| \mathbf{x} \|_1 = \sum | x_i |\ . \tag{4.76}$$

Norms have the following properties:

$$\| \mathbf{x} \| \geq 0 \quad \text{and} \ \| \mathbf{x} \| = 0 \Rightarrow \mathbf{x} = 0\,, \tag{4.77}$$

$$\| \alpha \mathbf{x} \| = | \alpha | \| \mathbf{x} \| \quad \text{for} \quad \alpha \in \mathbb{R}\,, \tag{4.78}$$

$$\| \mathbf{x} + \mathbf{y} \| \leq \| \mathbf{x} \| + \| \mathbf{y} \|\,. \tag{4.79}$$

4.4.1 General Stability Definitions

The splitting of the usually nonlinear equations of motion into a reference motion and into a perturbed motion was presented in Chapter 2. The reference motion itself might be a state of rest, but more frequently, it is some nonlinear type of motion, depending on the problem:

$$\mathbf{q}(t) = \mathbf{q}_0(t) + \eta_{\mathbf{q}}(t)\,. \tag{4.80}$$

With the norm of the perturbation vector $\eta_{\mathbf{q}}(t)$

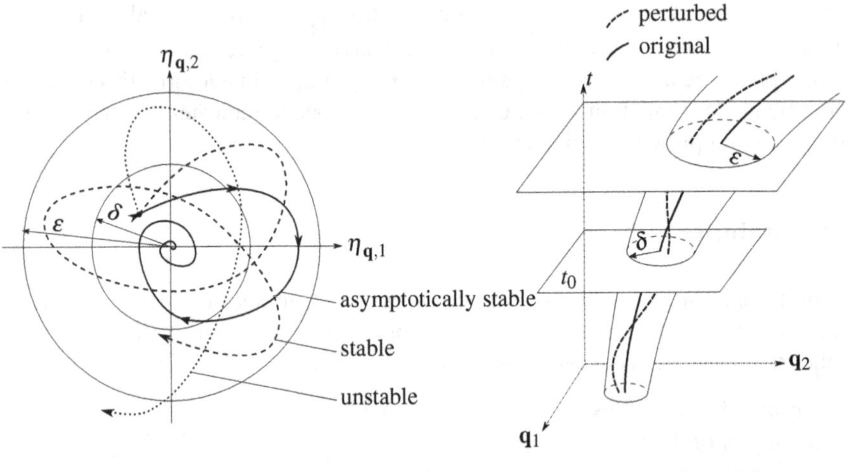

Fig. 4.12. Illustration of stability definitions: $\| \eta_{\mathbf{q}}(t_0) \| < \delta$ implies $\| \eta_{\mathbf{q}}(t) \| < \varepsilon$.

$$\| \eta_{\mathbf{q}} \| = \sqrt{\eta_{\mathbf{q}}^T \eta_{\mathbf{q}}}\,, \tag{4.81}$$

we are able to generalize the stability definition of LAGRANGE and DIRICHLET in Chapter 2 [36] (Fig. 4.12, left side):

Stability definition of LYAPUNOV:

- A unperturbed motion (state of rest) $\mathbf{q}_0$ of a dynamic system is called *stable*, if there exists for every real constant $\varepsilon > 0$ another real constant $\delta(\varepsilon) > 0$, so that from $\| \eta_{\mathbf{q}}(t_0) \| < \delta$, we always get $\| \eta_{\mathbf{q}}(t) \| < \varepsilon$ for all $t \geq t_0$.

- The unperturbed motion (state of rest) is called *asymptotically stable*, if it is stable and in addition $\lim_{t\to\infty} \| \eta_{\mathbf{q}}(t) \| = 0$.
- The unperturbed motion (state of rest) is called *stable in the limit*, if it is stable but not asymptotically stable.

Example 4.2 (Pendulum). Observation tells us that a pendulum (Fig. 4.13) has two states of rest, namely for $\varphi = 0$ and $\varphi = \pm\pi$. It is also well known that position $\varphi = 0$ will be stable, and position $\varphi = \pm\pi$ will be unstable. A small perturbation of position $\varphi = \pm\pi$ will cause the pendulum to start moving. These two pendulum states possess some typical features, which we consider in the following. The equations of motion write

$$\ddot{\varphi} + \omega^2 \sin\varphi = 0 \quad \text{with} \quad \omega^2 = \frac{mgl}{J}, \tag{4.82}$$

with the moment of inertia J about the point of suspension. Eliminating time by $\ddot{\varphi} = (\frac{d\dot{\varphi}}{d\varphi})\dot{\varphi}$ gives us

$$\dot{\varphi}\mathrm{d}\dot{\varphi} + \omega^2 \sin\varphi \mathrm{d}\varphi = 0 \tag{4.83}$$

and after integration the energy equation

$$\frac{1}{2}\dot{\varphi}^2 - \omega^2 \cos\varphi = c_1 . \tag{4.84}$$

Considering only small perturbations η_φ around the two states of rest, we may perform a TAYLOR-series expansion for (4.82) and (4.84):

- lower state of rest $\varphi_0 = 0$, $\varphi = \eta_\varphi$:

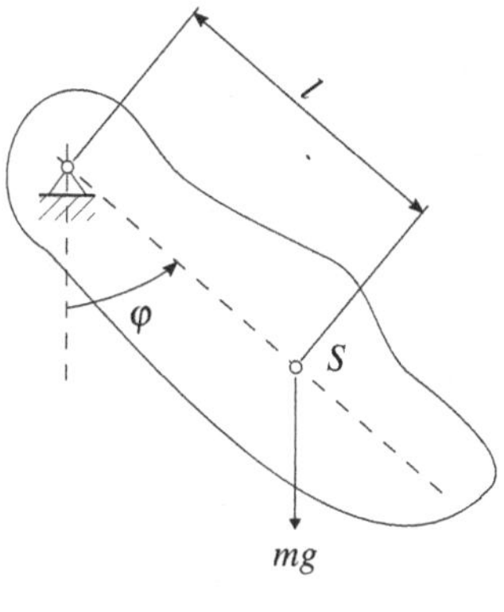

Fig. 4.13. Pendulum.

$$\ddot{\eta}_\varphi + \omega^2 \eta_\varphi = 0\,, \tag{4.85}$$
$$\dot{\eta}_\varphi^2 + \omega^2 \eta_\varphi^2 = c_2\,. \tag{4.86}$$

- upper state of rest $\varphi_0 = \pm\pi$, $\varphi = \pm\pi + \eta_\varphi$:

$$\ddot{\eta}_\varphi - \omega^2 \eta_\varphi = 0\,, \tag{4.87}$$
$$\dot{\eta}_\varphi^2 - \omega^2 \eta_\varphi^2 = c_2\,. \tag{4.88}$$

The solution of the state $\varphi_0 = 0$ describes an oscillation with constant amplitude

$$\eta_\varphi = \eta_{\varphi_0} e^{j\omega t}\,. \tag{4.89}$$

According to LYAPUNOV such an oscillation is stable in the limit. Equation (4.86) describes an ellipse in the phase plane $(\varphi, \dot{\varphi})$. Such points of equilibrium are said to be *elliptic points*. They are always stable.

The solution for the case $\varphi_0 = \pm\pi$ results in an oscillation with increasing amplitude

$$\eta_\varphi = \eta_{\varphi_0} e^{\omega t}\,. \tag{4.90}$$

The stability conditions of LYAPUNOV are not satisfied. The pendulum is unstable for that case. Equation (4.88) describes a hyperbola in the phase plane $(\varphi, \dot{\varphi})$. Such points are *hyperbolic points*. They are always unstable. Figure 4.14 displays the phase curves of all possible energy levels of the physical pendulum. It illustrates the areas of the stable elliptic points $\varphi_0 = 2\pi\nu$ as well as the unstable hyperbolic points $\varphi_0 = \pm(2\nu + 1)\pi$. The curves being on the top of these influence areas represent such a large energy that the pendulum starts to rotate. These two areas are

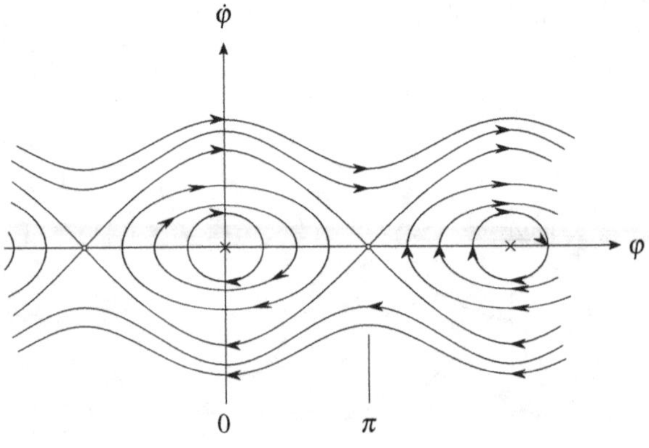

Fig. 4.14. Phase portrait for the pendulum.

separated by curves intersecting the unstable points $\varphi_0 = \pm(2\nu+1)\pi$. They form a *separatrix*.

The definitions of LYAPUNOV as given above refer to states of rest, so to say to a point in the phase space. If we want to refer these stability definitions to orbits, we recognize that the phase curves in the phase space change with time. Neighboring points stay close. Orbital stability requires that whole curves stay close without regarding specific points (right side of Fig. 4.12). We will come back to this point. See also the following example.

Example 4.3 (Orbital stability of a spherical pendulum.). We consider Fig. 4.15 and Example 2.1. A spherical pendulum will be stable with respect to a small perturbation of ϑ:

$$| \vartheta - \vartheta_0 | < \varepsilon_\vartheta \,.$$

However, a ε_ψ with $| \psi - \psi_0 | < \varepsilon_\psi$ for all $t \geq t_0$ does not exist, because $\dot{\psi}_0(t)$ represents a time-dependent reference motion. Anyway, the perturbed orbit will stay in the neighborhood of the reference curve, the pendulum will be orbitally stable.

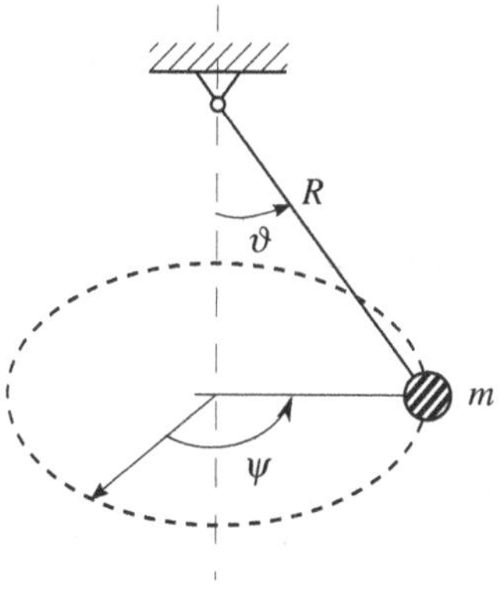

Fig. 4.15. Spherical pendulum.

4.4.2 Linear Stability

LYAPUNOV*'s first method* judges the stability of a nonlinear system from the linear terms of a TAYLOR series expansion of the equations of motion. Therefore, it is often called stability in the first approximation. LYAPUNOV formulated three stability conditions [10, 36].

We start with the nonlinear equations of motion in the state space form

$$\dot{\mathbf{x}} = \mathbf{f}(\mathbf{x}, t), \tag{4.91}$$

and we assume that $\mathbf{f}(\mathbf{0},t) = \mathbf{0}$ represents a state of rest. The term $\mathbf{h}(x,t)$ of the TAYLOR expansion

$$\dot{\mathbf{x}} = \mathbf{f}(\mathbf{x},t) = \underbrace{\mathbf{f}(\mathbf{0},t)}_{=\mathbf{0}} + \mathbf{A}(t)\mathbf{x} + \mathbf{h}(\mathbf{x},t) \tag{4.92}$$

is a vector-valued function, which does not include any linear components. Developing the nonlinear function $\mathbf{f}(\mathbf{x},t)$ makes sense, because the stability of a reference motion is usually perturbed by only small deviations. With the restriction to autonomous, time-invariant systems

$$\dot{\mathbf{x}} = \mathbf{A}\mathbf{x} + \mathbf{h}(\mathbf{x}) \,, \tag{4.93}$$

we formulate the following

Stability theorems of LYAPUNOV:

- If all eigenvalues λ_i of $\mathbf{A}$ have negative real parts, then the state of rest $\mathbf{x} = \mathbf{0}$ is *asymptotically stable*, independent of $\mathbf{h}(x)$.
- If at least one eigenvalue λ_k of $\mathbf{A}$ has a positive real part, then the state of rest $\mathbf{x} = \mathbf{0}$ is *unstable*, independent of $\mathbf{h}(x)$.

These two statements justify the limitation on investigations using only the linear term of a TAYLOR approximation. They also hold for the case of multiple eigenvalues. Some care is necessary, if the requirements for the above two theorems are not met:

- If the matrix $\mathbf{A}$ has no eigenvalues with positive real parts but with vanishing real parts, then the stability behavior is determined by the nonlinear term $\mathbf{h}(x)$.

An important question in connection with these theorems is the relation with the properties of the eigenvalues of $\mathbf{A}$. It can be answered applying the methods of Section 2.4. For the nonlinear case with vanishing real parts various methods are available, for example the center manifold reduction [24] or the general stability theory of LYAPUNOV.

4.4.3 Stability of Nonlinear Systems

For critical cases like the third theorem of the preceding section or for a form of the equations of motion, which does not allow a decomposition into linear and nonlinear parts, we have to determine stability only by considering the equations of motion directly. The crucial point is, that we want to achieve stability information without being forced to solve the equations of motion completely. LYAPUNOV considered and solved this problem in the year 1892 by developing his *second (direct) method* for the stability of motion [27]. He introduced a test function $V(\mathbf{x})$, which possesses some similarity with the energy of the system. This test function has to satisfy certain conditions thus giving statements of stability or no stability. Today, the test function V is called the LYAPUNOV function.

Before considering the geometry of such a test function more closely, we remind of the stability theorems of LAGRANGE and DIRICHLET from Section 2.4. These theorems connect stability of a state of rest with the minimum of the system's potential energy. In many cases, but definitely not in all cases, it is sufficient to choose the total energy for the LYAPUNOV function. A good choice is still a question of insight and experience.

To achieve a good feeling for the present method, we consider $V(\mathbf{x})$ as an n-dimensional cup in state space $\mathbf{x} \in \mathrm{IR}^{2f}$. This cup must satisfy certain features in connection with the equations of motion, to be able to deliver information on stability (Fig. 4.16). We restrict ourselves again to an autonomous system

$$\dot{\mathbf{x}} = \mathbf{f}(\mathbf{x}) \tag{4.94}$$

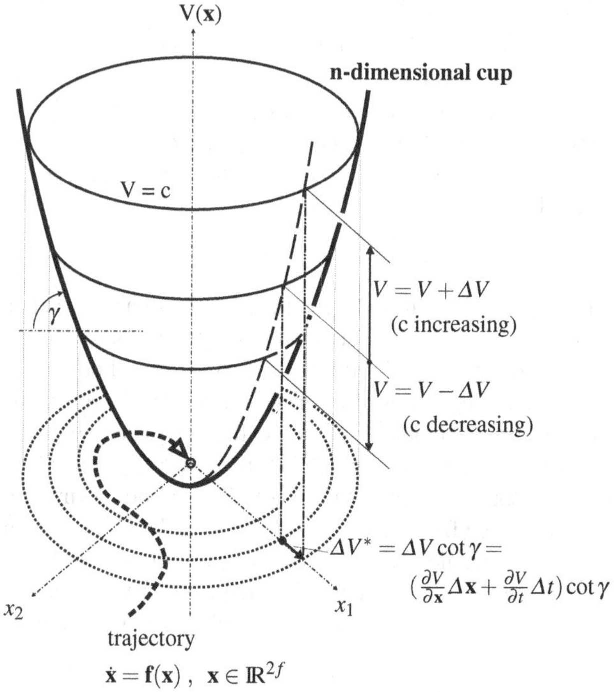

Fig. 4.16. n-dimensional cup $V(\mathbf{x})$ in state space.

with a state of rest $\mathbf{f}(\mathbf{0}) = \mathbf{0}$. A nonautonomous system $\dot{\mathbf{x}} = \mathbf{f}(\mathbf{x},t)$ would define a deformed and with time moving cup. The solution to the nonlinear equations of motion is an orbit in state space, which we do not know, and which we do not argue about. The cup picture includes the property, that the contour lines $(V = c)$ of the cup appear as closed curves on the cup surface. In addition, they might be projected onto the state space, which is indicated as a plane in Fig. 4.16.

With this picture in mind we are able to define a few conditions for $V(\mathbf{x})$ to be a LYAPUNOV function and the dynamic system to be stable:

1. $V(\mathbf{x})$, $\left(\frac{\partial V}{\partial \mathbf{x}}\right)$ are continuous at the origin.
2. $V(\mathbf{0}) = 0$ is an isolated minimum.
3. $V(\mathbf{x}) > 0$ around the origin (V is positive definite).
4. The derivation with respect to time along the solution trajectories $\dot{\mathbf{x}} = \mathbf{f}(\mathbf{x})$ satisfies

$$\left(\frac{\mathrm{d}V}{\mathrm{d}t}\right) = \frac{\partial V}{\partial \mathbf{x}}\frac{\mathrm{d}\mathbf{x}}{\mathrm{d}t} = \frac{\partial V}{\partial \mathbf{x}}\dot{\mathbf{x}} = \frac{\partial V}{\partial \mathbf{x}}\mathbf{f}(\mathbf{x}) \leq 0\,. \tag{4.95}$$

These features are immediately plausible in connection with the

stability propositions of LYAPUNOV:

- If we find a smooth positive definite function $V(\mathbf{x})$ with $(\frac{\mathrm{d}V}{\mathrm{d}t}) \leq 0$ in the neighborhood of $\mathbf{0}$, then $\mathbf{0}$ is *stable.*
- If we find a smooth positive definite function $V(\mathbf{x})$ with $(\frac{\mathrm{d}V}{\mathrm{d}t}) < 0$ in the neighborhood of $\mathbf{0}$, then $\mathbf{0}$ is *asymptotically stable.*
- If we find a smooth positive definite function $V(\mathbf{x})$ with $(\frac{\mathrm{d}V}{\mathrm{d}t}) > 0$ in the neighborhood of $\mathbf{0}$, then $\mathbf{0}$ is *unstable.*

As already indicated, the contour lines of the cup $V(\mathbf{x})$ are closed curves on the cup surface. Projecting these closed curves into the state space, which is displayed as a plane in Fig. 4.16, again results in closed curves around point $\mathbf{x} = 0$. If in the state space, the solution trajectories of $\dot{\mathbf{x}} = \mathbf{f}(\mathbf{x})$ run in such a way, that they cross the projected contour curves from outside to inside, then they are going to the reference state $\mathbf{x} = \mathbf{0}$ and represent stable solutions $(\dot{V} < 0)$. If such a projected contour line is itself a solution trajectory, then the solution is stable, but not asymptotically stable. A solution trajectory going from inside to outside with $(\dot{V} > 0)$ represents an unstable motion. This is the meaning of the above four conditions and the three stability propositions. Applying these rules, a good choice of the LYAPUNOV function V is mandatory. There exist no clear rules for such a choice, but certain examples and experiences:

1. We could choose the total energy and try to extend it by suitable terms.
2. Sometimes, we may find some known integrals of the equations of motion and use them for the LYAPUNOV function [48, 71].

- For conservative systems, we know the energy integral

$$T(\mathbf{q},\dot{\mathbf{q}})+V(\mathbf{q})=E_0\,. \tag{4.96}$$

- Again for conservative systems some momentum integrals may exist for the case that $\mathbf{q}_s$ does not appear in some terms for the energy (see Example 1.11). From the LAGRANGE's equations of the second kind, we get for example

$$\left(\frac{\partial T}{\partial \mathbf{q}}\right)_{\mathbf{q}_s}=\mathbf{0}\,,\quad \left(\frac{\partial V}{\partial \mathbf{q}}\right)_{\mathbf{q}_s}=\mathbf{0} \tag{4.97}$$

  and from this

$$\frac{d}{dt}\left(\frac{\partial T}{\partial \dot{\mathbf{q}}}\right)_{\mathbf{q}_s}=\mathbf{0}\,, \tag{4.98}$$

  which means

$$\left(\frac{\partial T}{\partial \dot{\mathbf{q}}}\right)_{\mathbf{q}_s}=\mathbf{p}_s^T=\text{const.} \tag{4.99}$$

  The magnitude $\mathbf{q}_s$ is said to be a cyclic coordinate and $\mathbf{p}_s$ a cyclic momentum. It is

$$\left(\frac{\partial T}{\partial \dot{\mathbf{q}}}\right)^T=\mathbf{p} \tag{4.100}$$

  a generalized momentum. It is a constant for cyclic coordinates and well suited as a possible part of a LYAPUNOV function.

In spite of the difficulties involved in finding a good LYAPUNOV function, at least in some cases, it should be kept in mind, that LYAPUNOV's methods are not only very ingenious, but also offer a quick possibility for a substantial stability analysis, even for complicated systems. We consider some simple examples in the following.

Example 4.4 (Nonlinear restoring forces and a critical case). We consider a motion with cubic restoring forces:

$$\dot{\mathbf{x}}=\begin{pmatrix}\dot{x}_1\\ \dot{x}_2\end{pmatrix}=\begin{pmatrix}0 & 1\\ -1 & 0\end{pmatrix}\begin{pmatrix}x_1\\ x_2\end{pmatrix}+a\begin{pmatrix}x_1^3\\ x_2^3\end{pmatrix}.$$

With respect to the state of rest $\mathbf{x}=\mathbf{0}$, this is a *critical case*, because the linear terms alone result in a characteristic equation $\lambda^2+1=0$, that is $\lambda=\pm j$ with $(\Re(\lambda)=0)$. Again for the linear terms alone the energy budget writes

$$x_1^2+x_2^2=c^2\,.$$

Therefore, we select a test function

$$V(\mathbf{x}) = x_1^2 + x_2^2$$

with the following properties:

- $V(\mathbf{0}) = 0$.
- $V(\mathbf{x}) > 0$ around the origin (V is positive definite).
- The derivation along the solution trajectory with respect to time gives

$$\frac{\mathrm{d}V}{\mathrm{d}t} = \frac{\partial V}{\partial x_1}\dot{x}_1 + \frac{\partial V}{\partial x_2}\dot{x}_2 = 2x_1\left(x_2 + ax_1^3\right) + 2x_2\left(-x_1 + ax_2^3\right) = 2a\left(x_1^4 + x_2^4\right) .$$

These features make V a LYAPUNOV function for $a \leq 0$:

$$a = 0 \Rightarrow \frac{\mathrm{d}V}{\mathrm{d}t} = 0 \quad \text{stable (stable at the limit)} ,$$
$$a < 0 \Rightarrow \frac{\mathrm{d}V}{\mathrm{d}t} < 0 \quad \text{asymptotically stable (damping)} .$$

For $a > 0$ the derivative satisfies $\frac{\mathrm{d}V}{\mathrm{d}t} > 0$ and the state of rest is unstable (excitation).

Example 4.5 (Center). We consider again a 1-DOF-oscillator with a nonlinear restoring force (4.12), which is assumed to behave near the origin $r(0) = 0$ like a strictly monotonic straight line. Nonlinear springs of some clutches behave like that. According to (4.12), we get

$$\ddot{x} + \frac{r(x)}{m} = 0 .$$

Expressing the acceleration by velocity, eliminating time, and separating the variables give

$$\dot{x}\mathrm{d}\dot{x} + \frac{r(x)}{m}\mathrm{d}x = 0 .$$

Integrating this equation with $\mathbf{x} = \left(x_1 \; x_2\right)^T = \left(x \; \dot{x}\right)^T$ around $\mathbf{x} = \mathbf{0}$ results in the energy equation

$$V(\mathbf{x}) = \frac{\dot{x}^2}{2} + \frac{1}{m}\int_0^x r(\xi)\mathrm{d}\xi = c^2 .$$

The first term corresponds to the kinetic, the second term to the potential energy. Choosing the energy $V(\mathbf{x})$ as a test function satisfies the conditions for a LYAPUNOV function, because we assume $xr(x) > 0$ for $x \neq 0$ and $r(0) = 0$. It is

$$\frac{\partial V}{\partial \mathbf{x}}\dot{\mathbf{x}} = \dot{x}_1\frac{\partial V}{\partial x_1} + \dot{x}_2\frac{\partial V}{\partial x_2} = x_2\frac{1}{m}r(x_1) - \frac{1}{m}r(x_1)x_2 = 0 .$$

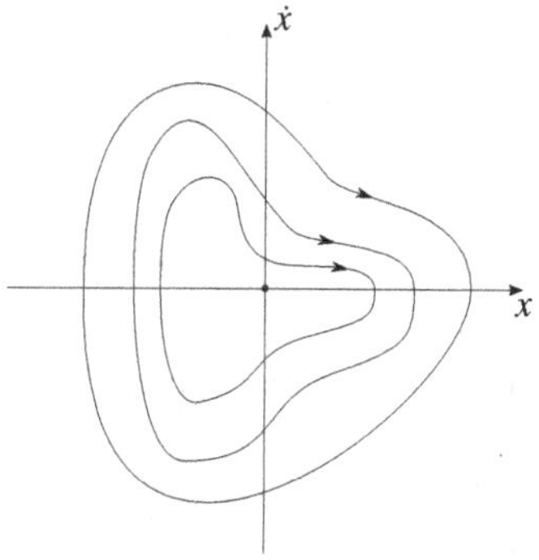

Fig. 4.17. Center point.

Therefore, the curves $V(\mathbf{x}) = c^2$ are closed curves, and the state of rest is stable in the limit (Fig. 4.17). The point with respect to the state of rest $\mathbf{x} = \mathbf{0}$ is also said to be a *center point*.

Example 4.6 (*Nondegenerated singularities* for the planar case). A nonlinear 1-DOF-oscillator can be described in state space by

$$\dot{\mathbf{x}} = \mathbf{f}(\mathbf{x}) = \mathbf{f}(\mathbf{0}) + \mathbf{A}\mathbf{x} + o\left(\| \mathbf{x} \|^2\right)$$

with

$$\mathbf{x} \in \mathrm{I\!R}^2 , \quad \mathbf{A} := \left(\frac{\partial \mathbf{f}}{\partial \mathbf{x}}\right)_{\mathbf{0}} = \begin{pmatrix} a & b \\ c & d \end{pmatrix} \in \mathrm{I\!R}^{2,2} , \quad \det(\mathbf{A}) \neq 0 .$$

For a state of rest $\mathbf{x} = \mathbf{0}$ (*singular point, equilibrium point*), we have $\mathbf{f}(\mathbf{0}) = \mathbf{0}$ and from this

$$\dot{\mathbf{x}} = \mathbf{A}\mathbf{x} .$$

The corresponding eigenvalues are

$$\lambda_{1,2} = \frac{1}{2}\operatorname{tr}(\mathbf{A}) \pm \frac{1}{2}\sqrt{\operatorname{tr}(\mathbf{A})^2 - 4\det(\mathbf{A})} = \frac{1}{2}(a+d) \pm \frac{1}{2}\sqrt{(a-d)^2 + 4bc}$$

with the eigenvectors

$$\mathbf{x}_1 = \begin{pmatrix} 1 \\ -\left(\frac{a-\lambda_1}{b}\right) \end{pmatrix} , \quad \mathbf{x}_2 = \begin{pmatrix} 1 \\ -\left(\frac{a-\lambda_2}{b}\right) \end{pmatrix} .$$

From this, the general solution with arbitrary constants c_1 and c_2 writes

$$\mathbf{x}(t) = c_1\mathbf{x}_1 e^{\lambda_1 t} + c_2\mathbf{x}_2 e^{\lambda_2 t} ,$$
$$\dot{\mathbf{x}}(t) = c_1\lambda_1\mathbf{x}_1 e^{\lambda_1 t} + c_2\lambda_2\mathbf{x}_2 e^{\lambda_2 t} .$$

According to the solution behavior around the state of rest, we describe a couple of variants by the various eigenvalue possibilities of the matrix $\mathbf{A}$.

1. ***Saddle point*** (2-tangent node)
   λ_1 and λ_2 have opposite signs and are *real*, which means

$$\det(\mathbf{A}) = \lambda_1\lambda_2 = ad - bc < 0 ,$$
$$\Delta = \mathrm{tr}(\mathbf{A})^2 - 4\det(\mathbf{A}) = (a-d)^2 + 4bc > 0 .$$

   One part of the solution goes in the direction of the equilibrium point, the other part goes the opposite way (Fig. 4.18).

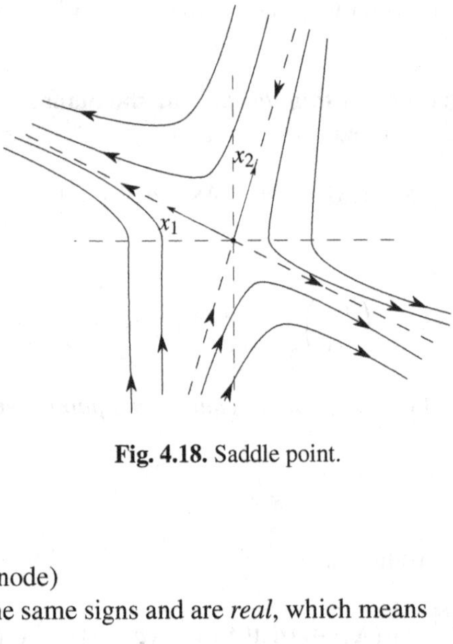

Fig. 4.18. Saddle point.

2. ***Node*** (2-tangent node)
   λ_1 and λ_2 have the same signs and are *real*, which means

$$\det(\mathbf{A}) = \lambda_1\lambda_2 = ad - bc > 0 ,$$
$$\Delta = \mathrm{tr}(\mathbf{A})^2 - 4\det(\mathbf{A}) = (a-d)^2 + 4bc > 0 .$$

   The solutions go in the direction of the equilibrium point or they go the opposite way depending on the sign of $\mathrm{tr}(\mathbf{A})$ (Fig. 4.19).

3. ***Focus / vortex***
   λ_1 and λ_2 are conjugate complex (and therefore also the eigenvectors $\mathbf{x}_1$ and $\mathbf{x}_2$), which means

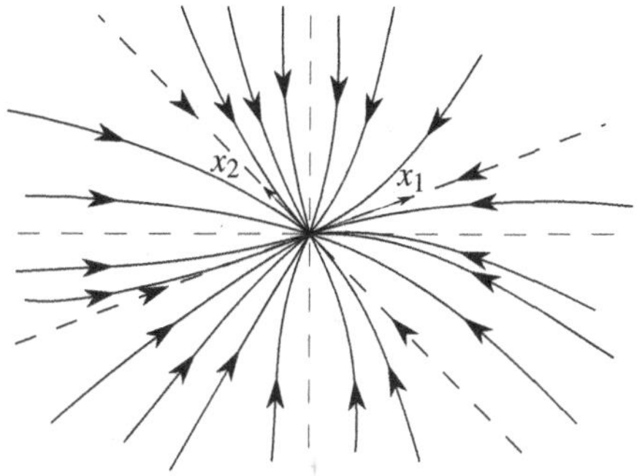

Fig. 4.19. Stable node ($\mathrm{tr}(\mathbf{A}) < 0$).

$$\det(\mathbf{A}) = \lambda_1\lambda_2 = ad - bc > 0\,,$$
$$\Delta = \mathrm{tr}(\mathbf{A})^2 - 4\det(\mathbf{A}) = (a-d)^2 + 4bc < 0\,.$$

We assume $\mathrm{tr}(\mathbf{A}) \neq 0$. The solutions go to the origin or they go away from the origin, according to the sign of $\mathrm{tr}(\mathbf{A})$ (Fig. 4.20).

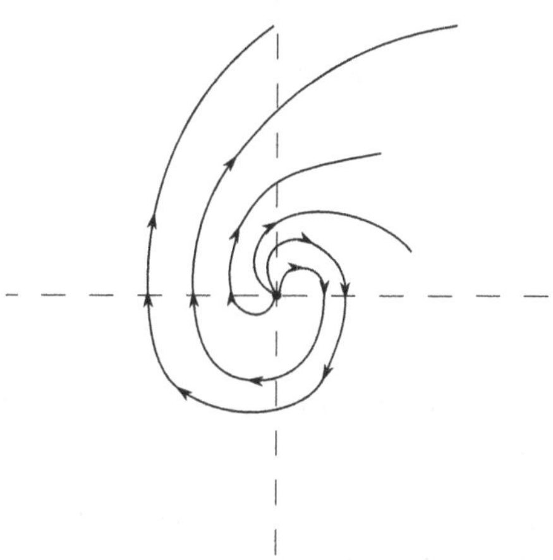

Fig. 4.20. Unstable node (vortex).

4. ***Center point***
λ_1 and λ_2 are purely imaginary, which means, that similar to the node

$$\det(\mathbf{A}) = \lambda_1\lambda_2 = ad - bc > 0\,,$$
$$\Delta = \mathrm{tr}(\mathbf{A})^2 - 4\det(\mathbf{A}) = (a-d)^2 + 4bc < 0\,.$$

Because of $\mathrm{tr}(\mathbf{A}) = 0$, we get a center (vortex) comparable to Example 4.6 (Fig. 4.21).

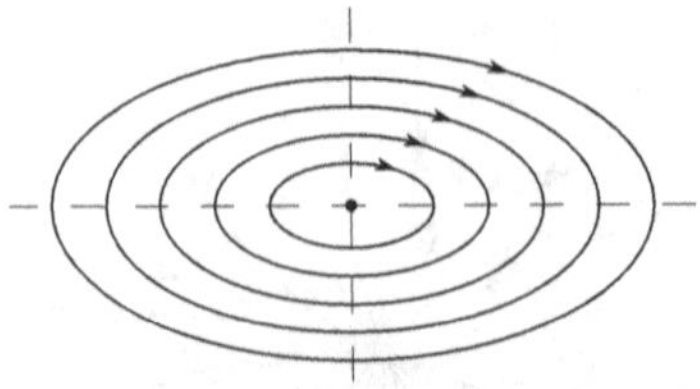

Fig. 4.21. Center (vortex).

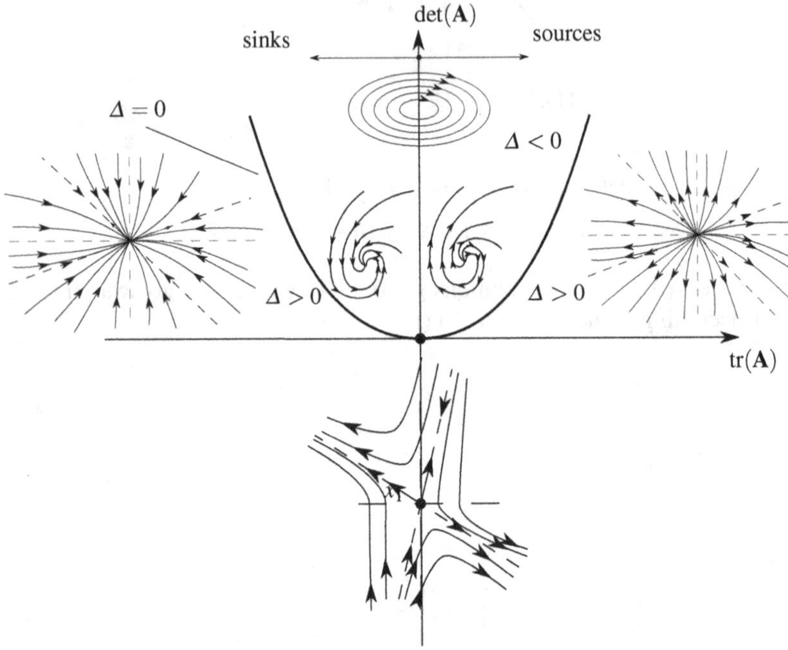

Fig. 4.22. Combined diagram ($\Delta = \mathrm{tr}(\mathbf{A})^2 - 4\det(\mathbf{A})$).

It is possible to combine all these cases in one diagram (Fig. 4.22). We see that for cases 1, 2, and 3, the nonlinear system behaves approximately like a linear system in the neighborhood of the singularity. This is not true for case 4. The nonlinear system of case 4 may also have a node (vortex), which finally is decided by terms of higher order in the Taylor series of **f** or by the general method of LYAPUNOV according to Example 4.6. Such cases are called *degenerated singularities* or *singularities of higher order*. The same situation might occur if at least one of the eigenvalues λ_1 or λ_2 is a null-eigenvalue or if the matrix **A** cannot be diagonalized. The above-mentioned classification of singularities was already established by POINCARÉ more than a hundred years ago [24], whereas these degenerated singularities are still a matter of research today, especially with respect to singularities in multidimensional systems.

Chapter 5
Vibration Phenomena

5.1 Introduction

As we have seen in previous chapters, purely periodic oscillations can be characterized by a periodicity condition

$$\mathbf{x}(t) = \mathbf{x}(t+T) \tag{5.1}$$

with the period T and the frequency

$$f = \frac{1}{T} \tag{5.2}$$

or the angular frequency

$$\omega = 2\pi f = 2\frac{\pi}{T} \,. \tag{5.3}$$

The period need not be constant, but may depend on the amplitude (Fig. 4.7). In the following, we deal with vibrations with one or more constant periods and investigate the issue of vibration formation. In order to recognize the essentials, we restrict ourselves in many cases to oscillators with one degree of freedom $x(t)$.

Oscillations are characterized by periodically recurring states $(x, \dot{x})$, where the intensity of these states may change depending on the energy in the system. This is reflected by the oscillation amplitude, which may be damped by dissipation, or also be excited. The recurring states can be realized by different physical mechanisms.

The simplest cases are *free* and *forced oscillations* as described in Chapters 2 and 3. Characteristic of these oscillations are the eigenfrequencies f or the excitation frequencies Ω. While modes are created by a single external impulse, forced vibrations require an (usually periodic) external excitation. Forced oscillations are the cause behind most problems in machine dynamics.

As long as vibrations can be described by linear differential equations with constant coefficients, a more or less closed mathematical tool box to solve the problems is available [1, 46]. However, some important oscillation phenomena are highly

F. Pfeiffer and T. Schindler, *Introduction to Dynamics*,
DOI: 10.1007/978-3-662-46721-3_5

nonlinear. Their mathematical treatment is far more difficult. These include *self-excited* and *parametrically excited oscillations*. Self-excited vibrations are created as a dynamic equilibrium state of power supply into the system and energy consumption within the system, where the energy is supplied at the frequency of the vibrations. Examples are flutter, friction oscillators, and the pendulum clock. For parametrically excited oscillations, the periodic changes of one or more parameters of the system work as a kind of internal excitation, for example, the periodically variable tooth stiffness due to the time-varying tooth contact in a gear [51, 42].

The frequencies of modes and self-excited oscillations are real eigenfrequencies, because they are determined by the vibration system itself (autonomous systems). In contrast, the frequencies of forced and parametrically excited vibrations depend on external excitations or design-related time-varying processes (external excitation, heteronomous systems). Table 5.1 gives an overview of the oscillation phenomena to be treated. The classification of the vibrations according to *formation principles* is certainly useful, but not the only possibility. Vibrations can also be distinguished according to their *properties*, in linear and nonlinear oscillations. A third aspect could be the *degree of freedom*, that is the antiquated distinction between *mono-track* and *multiple-track* oscillators (*coupled vibrations*). Although most machines represent oscillators with many degrees of freedom, the restriction to one degree of freedom makes sense, because its characteristic behavior can also be found in large systems.

Table 5.1. Classification of vibrations according to formation principles.

vibration type	examples	origin	frequency	equation of motion
free vibrations	pendulum, tuning fork, piano string modes	single external impulse	eigen angular frequency ω	homogeneous $\ddot{x}+\omega^2 x=0$
forced vibrations	foundation vibration, vibrating screen, vehicles on rough ground	external forces or moments, usually acting periodically	excitation frequency Ω	inhomogeneous $\ddot{x}+\omega^2 x=\bar{F}\cos(\Omega t)$
self-excited vibrations	clock, bell, string and wind instruments, flutter, woodpecker toy	self-control with not periodically acting energy source	about eigen angular frequency ω	nonlinear $\ddot{x}+f(x,\dot{x})=0$
parameter-excited vibrations	piston engines, propeller, gear drive, pendulum with moving point of suspension	periodically changing parameters	fractions or multiples of the parameter frequency ω_P	periodic coefficients $\ddot{x}+p(t)x=0$

5.2 Free Vibrations

Free vibrations can be characterized by their eigenfrequencies, their eigenvectors or modes, their damping behavior and energy balance. For a conservative system, there is a periodic exchange of potential and kinetic energy (pendulum in Fig. 5.1):

$$T + V = E_0 = \text{const.} \tag{5.4}$$

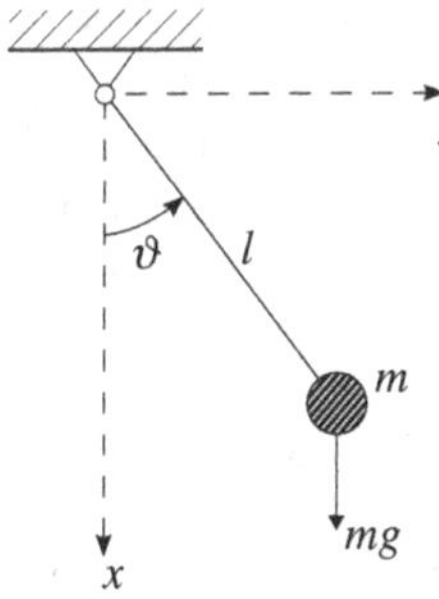

Fig. 5.1. Pendulum.

Modes may occur in linear and nonlinear systems. For a small initial deflection, the pendulum in Fig. 5.1 is an example of a linear mode:

$$\left(ml^2\right)\ddot{\vartheta} = -mgl\vartheta\,. \tag{5.5}$$

With $\omega^2 = \frac{g}{l}$, this yields

$$\ddot{\vartheta} + \omega^2\vartheta = 0\,. \tag{5.6}$$

The solution for this system is

$$\vartheta(t) = \vartheta_0\cos(\omega t) + \frac{\dot{\vartheta}_0}{\omega}\sin(\omega t)\,. \tag{5.7}$$

With (5.6), introducing the velocity, eliminating the time, and separating the variables gives

$$\dot{\vartheta}\mathrm{d}\dot{\vartheta} + \omega^2\vartheta\mathrm{d}\vartheta = 0\,. \tag{5.8}$$

In the phase portrait $\left(\vartheta, \dot{\vartheta}\right)$, we obtain the ellipses

$$\vartheta^2 + \left(\frac{\dot{\vartheta}}{\omega}\right)^2 = \vartheta_0^2 + \left(\frac{\dot{\vartheta}_0}{\omega}\right)^2\,. \tag{5.9}$$

If we also allow large deflections ϑ, we obtain the equation of motion

$$\ddot{\vartheta} + \omega^2 \sin \vartheta = 0 \tag{5.10}$$

and therefore the phase curves

$$\left(\frac{\dot{\vartheta}}{\omega}\right)^2 - 2\cos\vartheta = \left(\frac{\dot{\vartheta}_0}{\omega}\right)^2 - 2\cos\vartheta_0 \,, \tag{5.11}$$

which degenerate to (5.9) for small ϑ. The corresponding phase portrait is shown in Fig. 5.2, from which we get the periodic change of the potential energy V (Example 4.2). Another example of a nonlinear mode is given by the rod on a block in Fig. 4.12. Phase portrait and amplitude response are shown in Fig. 5.3.

For vibration processes with energy dissipation, the eigenfrequencies change only slightly, but the amplitude changes significantly. Considering the single degree of freedom oscillator in Fig. 5.4, the following equations can be stated:

$$m\ddot{x} + d\dot{x} + cx = F(t) = 0 \,, \tag{5.12}$$

or with the eigen angular frequency of the undamped system $\omega_0^2 = \frac{c}{m}$ and the LEHR *damping* $D = \frac{d}{2m\omega_0} = \frac{d}{2\sqrt{cm}}$:

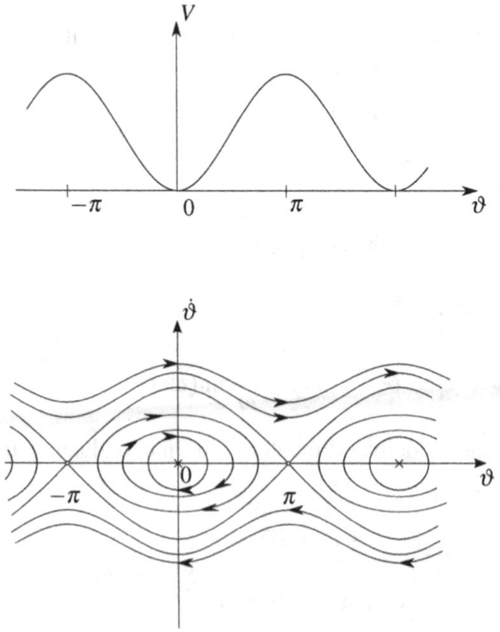

Fig. 5.2. Phase portrait of the pendulum.

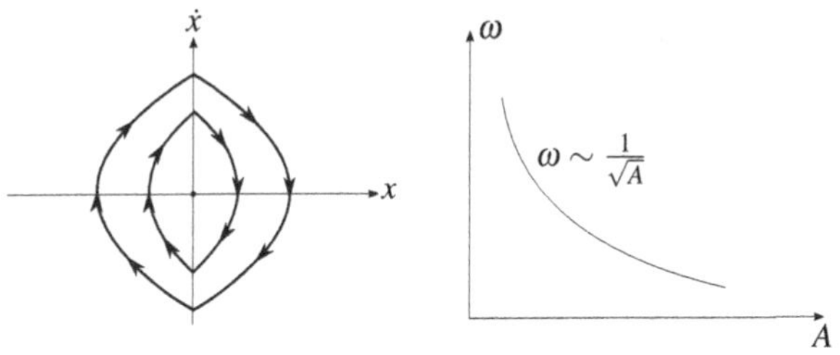

Fig. 5.3. Rod on a block – phase portrait and amplitude response (see Figs. 4.5 and 4.6).

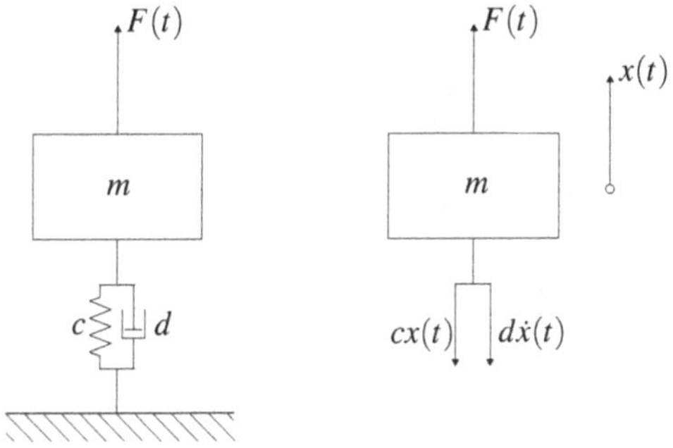

Fig. 5.4. Single degree of freedom oscillator.

$$\ddot{x} + 2D\omega_0\dot{x} + \omega_0^2 x = 0\,. \tag{5.13}$$

With the ansatz $x = \bar{x}e^{\lambda t}$, we obtain the eigenvalues

$$\lambda_{1,2} = -D\omega_0 \pm j\omega_0\sqrt{1-D^2}\,. \tag{5.14}$$

The classification of the vibration types is thus given as in Fig. 2.5a-2.5d. Depending on the value of D, five cases occur.

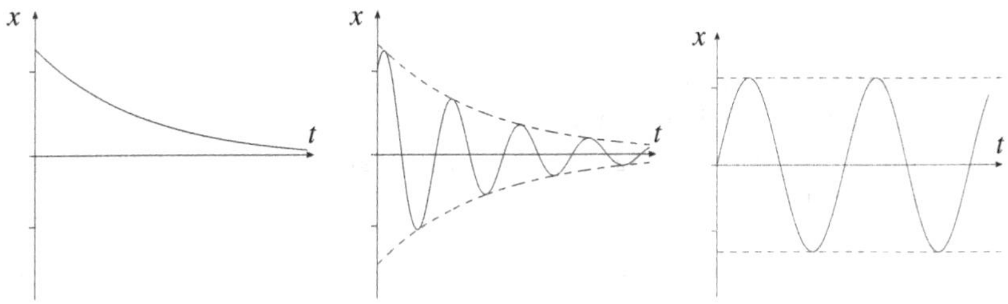

(a) Aperiodic case. (b) Damped vibration. (c) Periodic vibration.

Fig. 5.5. Stable oscillation types.

1. *Aperiodic case*, no oscillation ($D > 1$, Fig. 5.5a)

$$\lambda_{1,2} = \omega_0 \left(-D \pm \sqrt{D^2 - 1}\right) . \tag{5.15}$$

2. *Damped vibration* ($0 < D < 1$, Fig. 5.5b)

$$\lambda_{1,2} = \omega_0 \left(-D \pm j\sqrt{1 - D^2}\right) . \tag{5.16}$$

3. *Periodic vibration*, limiting case ($D = 0$, Fig. 5.5c)

$$\lambda_{1,2} = \pm j\omega_0 . \tag{5.17}$$

4. *Excited vibration* ($-1 < D < 0$, Fig. 5.6a)

$$\lambda_{1,2} = \omega_0 \left(-D \pm j\sqrt{1 - D^2}\right) . \tag{5.18}$$

5. *Unstable aperiodic case* ($D < -1$, Fig. 5.6b)

$$\lambda_{1,2} = \omega_0 \left(-D \pm \sqrt{D^2 - 1}\right) . \tag{5.19}$$

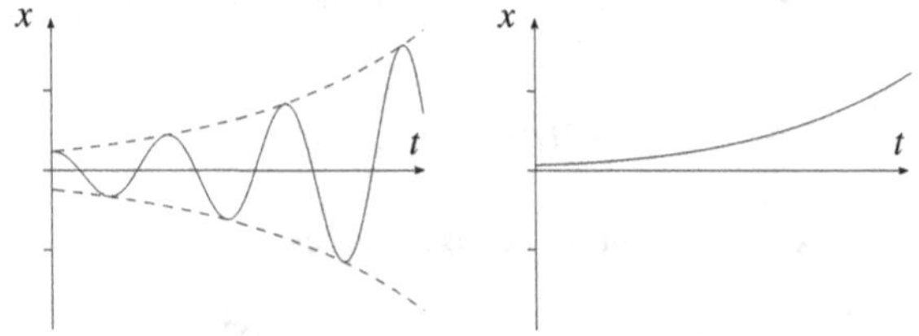

(a) Excited vibration. (b) Unstable aperiodic case.

Fig. 5.6. Unstable oscillation types.

5.3 Forced Vibrations

Forced vibrations are characterized by the eigen behavior of the vibration system, by the type of excitation, and the resulting output in the form of *amplitude-* and *phase frequency response functions* [42, 41]. We consider an oscillator with one degree of freedom as in (5.12) with right-hand side $f(t) \neq 0$

$$m\ddot{x} + d\dot{x} + cx = f(t) = c\bar{A}\cos(\Omega t) \tag{5.20}$$

resulting from a harmonic kinematic excitation $\bar{A}\cos(\Omega t)$. With the eigen angular frequency of the undamped system ω_0 and the LEHR damping, we obtain the differential equation:

$$\ddot{x} + 2D\omega_0\dot{x} + \omega_0^2 x = \frac{c\bar{A}}{m}\cos(\Omega t) . \tag{5.21}$$

The solution consists of a homogeneous and an inhomogeneous part. The homogeneous component corresponds formally to the solution of (5.13) and physically to the modes. If we assume a damped system, the modes initiated only once decrease and disappear after a while. In contrast, the continuously excited oscillations persist, because the periodic excitation supplies power and forces the vibration system to oscillate with the same frequency but different amplitude and phase.

According to Section 2.3.1, the stationary solution results from the frequency response function ($\omega_0^2 = \frac{c}{m}$, $D = \frac{d}{2m\omega_0} = \frac{d}{2\sqrt{cm}}$, $\eta = \frac{\Omega}{\omega_0}$)

$$\mathrm{G}(j\Omega) = \frac{c}{m\left(-\Omega^2 + j\Omega 2D\omega_0 + \omega_0^2\right)} = \frac{1}{1 + j2D\eta - \eta^2} . \tag{5.22}$$

Amplitude amplification and phase shift satisfy

$$V = |\mathrm{G}(j\Omega)| = \frac{1}{\sqrt{(1-\eta^2)^2 + 4D^2\eta^2}} , \tag{5.23}$$

$$\psi(\mathrm{G}(j\Omega)) = \arctan\left(\frac{2D\eta}{1-\eta^2}\right) , \tag{5.24}$$

such that the particular solution is

$$x = AV\cos(\Omega t - \psi) . \tag{5.25}$$

The ansatz *of the type of the right-hand side* is based on the physically reasonable assumption that the *input* $A\cos(\Omega t)$ creates an *output*, which has a phase shift ψ and an amplitude changed by the amplification factor V. The maximum value of V is derived from $\frac{\partial V}{\partial \eta} = 0$:

$$\eta_{\max} = \sqrt{1 - 2D^2} . \tag{5.26}$$

Plotting V and ψ over η, we obtain the amplitude frequency response function in the first case (Fig. 5.7) and the phase frequency response function (Fig. 5.8) in the second case.

Such diagrams are indispensable tools for the estimation of vibrations, especially in systems with many degrees of freedom. We can calculate and measure the amplitude and phase frequency response functions. The evaluation of complex systems requires carefully established models with clear relation to practice. Measurements

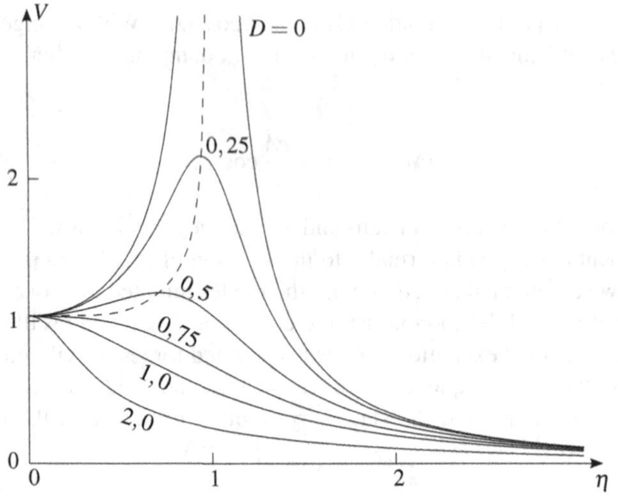

Fig. 5.7. Amplitude frequency response function (5.23).

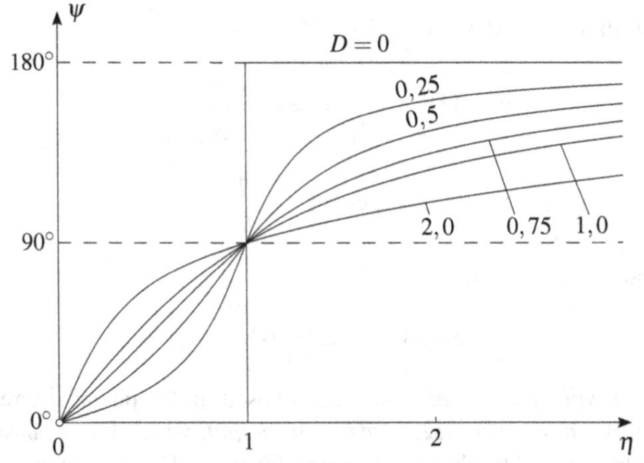

Fig. 5.8. Phase frequency response function (5.24).

demand care and a clever (meaningful) arrangement of the measuring points. In each case, the curves inform on

- resonance points,
- damping behavior,
- phase of the degrees of freedom with respect to each other,
- assessment of the resonances,
- impact of construction parameter variations on position and amplitude of resonances.

We discuss forced oscillations with the actual example of a ship propulsion. The five-bladed propeller is driven by a gas turbine via a planetary gear and via a spur gear (Fig. 5.9). The most important external excitation is caused by the propeller. It takes effect with five times the propeller speed (= output speed), since each propeller blade produces a perturbation of the mean propeller thrust due to the hydrodynamic processes at the ship's stern. Although the axial thrust bearing absorbs the total thrust and the majority of the thrust variations, the rotational nonuniformity of the propeller shaft as well as a part of the axial thrust variations remain and negatively influence the entire drive train.

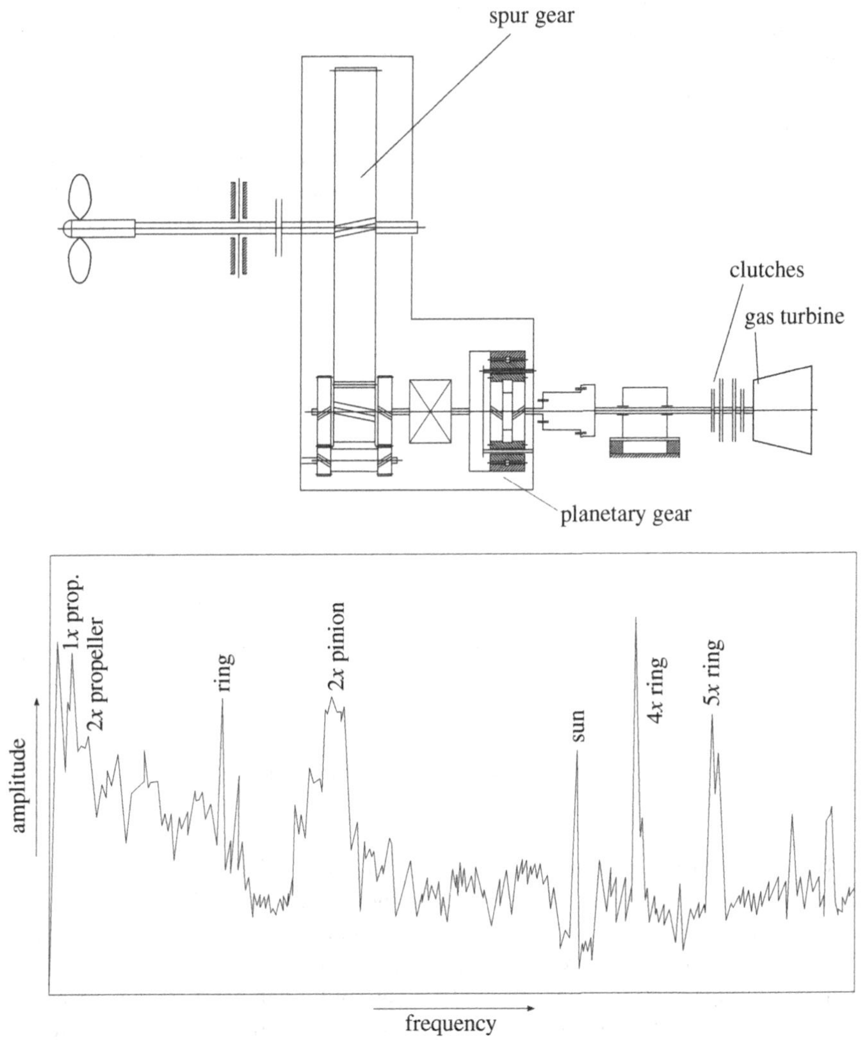

Fig. 5.9. Ship propulsion, measuring point located at the planetary gear.

In addition to the external excitations, there are parametric excitations in the toothing of the two gears. As there is never the same number of teeth in contact and the toothing effects the overall dynamics of the system as an elastic coupling, we obtain a time-varying, that is not constant, stiffness characteristic in such coupling points. The deviation from the constant mean depends on the degree of coverage of the gears (Section 5.5). The time-varying elastic couplings generate parametric excitations in the system that lead to additional resonances. These parametric excitations are usually locally limited to the environment of their origin, especially in large powertrains [35]. Since the spectrum in Fig. 5.9 has been recorded at the planetary gear, we can only observe the influence of parametric excitations at the planetary gear and not those at the spur gear.

The influence of the propeller excitation occurs only in the lower frequency range. At higher frequencies, the parametrically excited vibrations and the modes of all the components of the planetary gear dominate. These can be more dangerous with respect to the amplitude increase than the lower frequency resonance (Section 5.5). The amplitude frequency response function (*spectrum*) for the main harmonic components (FOURIER coefficients) of a vibration system does not contain direct information on the energy of the resonances. This topic is addressed in more detail in [42, 43].

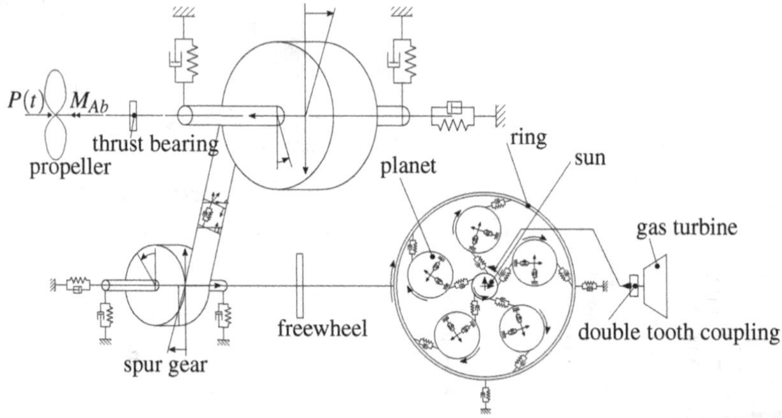

Fig. 5.10. Model of the ship propulsion (Fig. 5.9).

The analysis of a spectrum as in Fig. 5.9 may be difficult. Often, we have to find a mechanical vibration model. Fig. 5.10 shows such a model for the drivetrain depicted in Fig. 5.9. Basically, it serves to represent the torsional vibrations of the overall system and some more detailed phenomena in the gears. All coupling points such as bearings and toothings are modelled as spring-damper elements, whereby we have to take into account the time-varying stiffness and damping in the toothings (Section 5.5). For such a mechanical model, the equations of motion are derived according to the methods of Chapter 1. After that, simulations can be undertaken with the mathematical model. In the present case, the entire drivetrain system has 38 degrees of freedom. The measured frequencies in Fig. 5.9 could be identified and proven without exception.

5.4 Self-excited Vibrations

Self-excited vibrations represent a very special and fascinating class of natural oscillations. They take energy from an energy source at the frequency of the vibrations and cover energy losses in the vibration system with this energy supply. If there is a balance of supply and loss, a stable periodic oscillation occurs that can be represented by a *limit cycle* in the phase diagram. Self-excited oscillators need an energy source and a kind of *switch*, which can enable or disable the supply of energy from the *source*. Since the oscillation is periodic in its stationary state, the switch must be involved in the periodic process of the oscillator. For self-excited oscillators, we distinguish two types, that is *oscillator type* and *storage type*. A typical example of the oscillator type is the electric bell (Fig. 5.11). It takes electric energy from the source *grid* and thus causes an electromagnet to attract the clapper and hit the bell. This process interrupts the power supply such that the clapper swings back into its original position due to the deformation energy stored in the retaining spring. Then, the process begins again. Switch, energy storage, and oscillator is the clapper with its elastic attachment, while the energy source is the grid [42]. Other examples of oscillator-type self-excited oscillators are the pendulum clock, the violin string, flutter, wind instruments, and the KÁRMÁN vortex street. The violin string stands for a whole class of self-excited oscillators, the *friction oscillators.*

Self-excited oscillators of storage type are often relaxation oscillators on a hydraulic basis. A vessel is filled and emptied at a certain filling level through a tube (Fig. 5.12), or mechanical tilting occurs.

For the basic understanding of self-excited vibrations, we have to analyze the energy balance. Linear and nonlinear conservative oscillators have modes that are

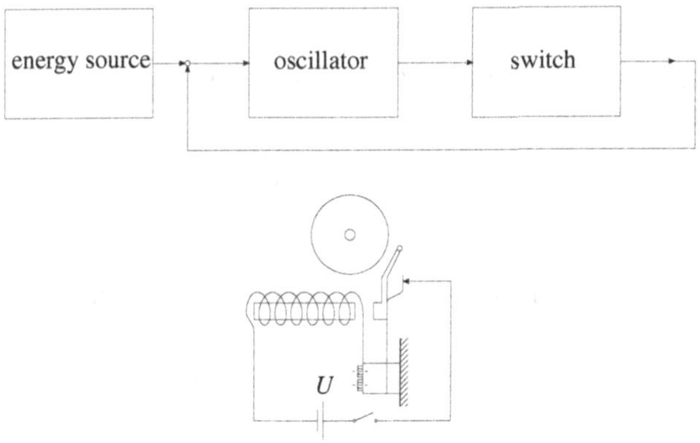

Fig. 5.11. Self-excited oscillator: oscillator type, example bell.

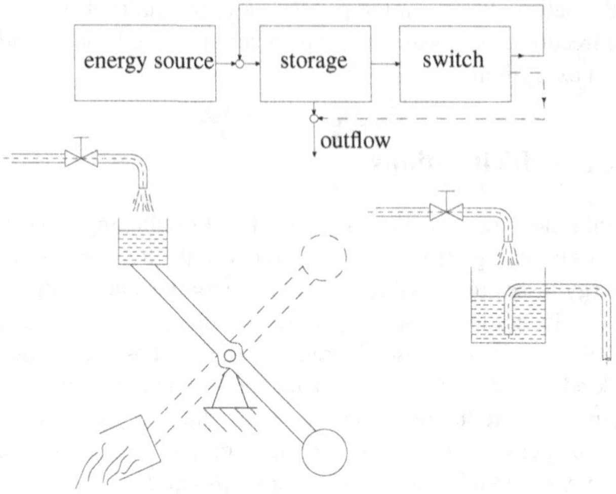

Fig. 5.12. Self-excited oscillator: storage type

characterized by a periodic exchange of potential and kinetic energy (5.4). The key energy magnitudes are T and V. In contrast, for self-excited vibrations the motion is determined by the supplied energy ΔE_Z and by the dissipated energy ΔE_D in the system. During a period, we have to supply the same amount of energy ($+\Delta E_Z$) as is dissipated in the system ($-\Delta E_D$). The following cases are possible:

- $\Delta E_Z < \Delta E_D$ (damping):
  More energy is dissipated than supplied. The amplitude decreases.
- $\Delta E_Z > \Delta E_D$ (excitation):
  More energy is supplied than dissipated. The amplitude increases.
- $\Delta E_Z = \Delta E_D$ (limit cycle):
  Just as much energy is supplied as is dissipated. This is the limiting case of periodic motion.

To get a rough quantitative estimate of these ratios, we assume a periodic oscillation and a linear damping law:

$$x \approx \bar{x}\cos(\omega t)\,, \tag{5.27}$$

$$F_D \approx -d\dot{x}\,. \tag{5.28}$$

Then for an oscillator with one degree of freedom, we get the dissipated energy

$$\Delta E_D \approx -\int_0^T F_D\dot{x}\mathrm{d}t = d\int_0^T \dot{x}^2\mathrm{d}t = \pi d\omega A^2 \sim A^2\,. \tag{5.29}$$

For simplicity, we consider an oscillator for which the supplied energy ΔE_Z is constant. Then, we obtain the conditions of Fig. 5.13, which are typical for self-excited

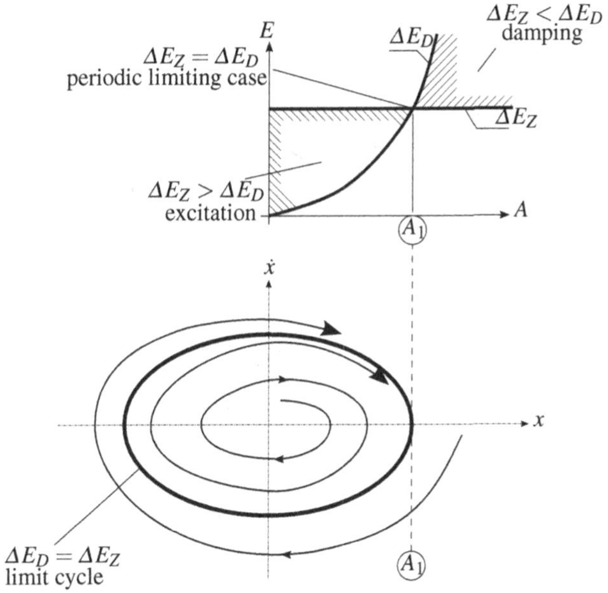

Fig. 5.13. Energy balance.

oscillators. The point with the same amount of supplied and of dissipated energy defines a stable limit cycle in the illustrated case, which separates the two regions of excitation and damping. In the phase portrait, this means that the vibrations approach the limit cycle from outside and from inside and remain stable on the limit cycle. There are also unstable limit cycles, for which the regions of excitation and damping are reversed, that is the damping region is below and the excitation region is above the limit cycle. We can observe such a case for the pendulum clock, for which the basic conditions are sketched in Fig. 5.14. The power supply is strongly amplitude-dependent, causing a second unstable limit cycle [33, 42]. If the pendulum is initiated only very weakly, the pendulum clock mechanism (for example, the GRAHAM mechanism in Section 5.4.7) does not start and the pendulum gets to rest due to internal friction. With moderate initiation, the pendulum clock mechanism is started; it drives the pendulum amplitudes to the outer stable limit cycle, which corresponds to the continuous operation of the clock.

5.4.1 Hydraulic Oscillator

Hydraulic oscillators belong to the storage type. The basic structure of such a system is shown in Fig. 5.12. For the hydraulic oscillator shown in Fig. 5.15, the storage is filled by a continuous stream of water. At the water level height of h_2, a lever attached in the storage, that is a switch acting on the sink, becomes active and ensures that the storage is emptied to the height h_1. Because of the air entering the system, the emptying is interrupted. Subsequently, it once again begins to fill. The

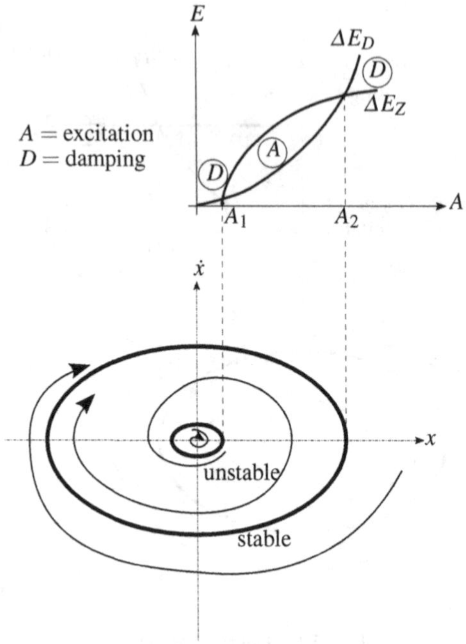

Fig. 5.14. Principle of the pendulum clock.

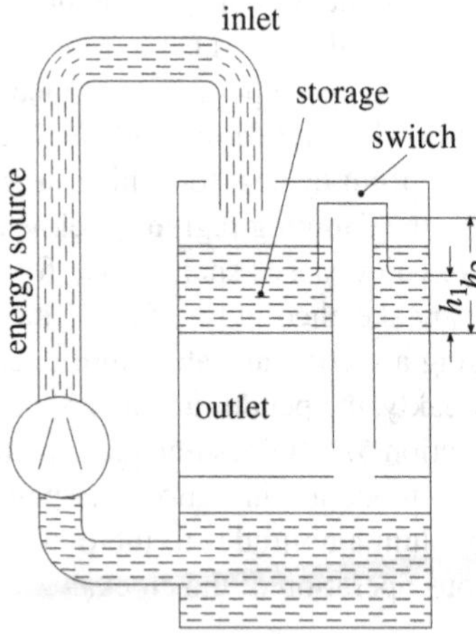

Fig. 5.15. Hydraulic oscillator.

motion consists of a repetitive swinging of the water height h between the two limiting heights h_1 and h_2, where the oscillation time T is the sum of filling time T_F and emptying time T_E. The time behavior of the system is shown in Fig 5.16a, while the limit cycle of the phase portrait belonging to the steady state is shown in Fig. 5.16b.

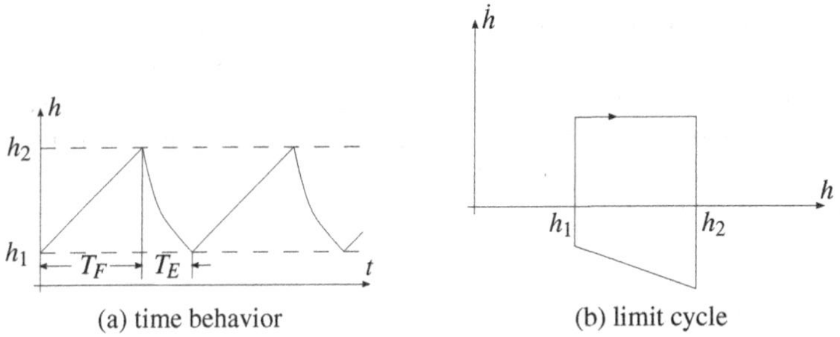

Fig. 5.16. Behavior of the hydraulic oscillator.

5.4.2 *Drinking Bird*

The well-known drinking bird toy (Fig. 5.17) is a self-excited system of storage-oscillator type. Apart from the energy source (environment) and the switch (riser pipe), both storage (head and body) and oscillator (pendulum arrangement) are necessary such that the thermal-mechanical system can perform self-excited oscillations. An absorbent layer fitted to the head of the drinking bird, which can soak water through the pecker, ensures that the head remains wet and cool. The neck is designed as a riser pipe, it extends into the body and is surrounded by ether or another low-boiling liquid. Both in the body and in the head, the steam of the ether is above the liquid. The vapor pressure p_1 in the body equals approximately the one of the environment. The pressure p_2 in the head is lower because the head temperature is about $0,3^\circ$ below the room temperature due to the water evaporation. Because of the pressure difference, the liquid level in the riser pipe increases (height h). The center of gravity *S*, which was initially below the pivot point *D*, shifts upwards and the drinking bird tilts forwards slowly. In the nearly horizontal position, the pecker dips into the water. The amount of liquid in the interior of the bird is defined, such

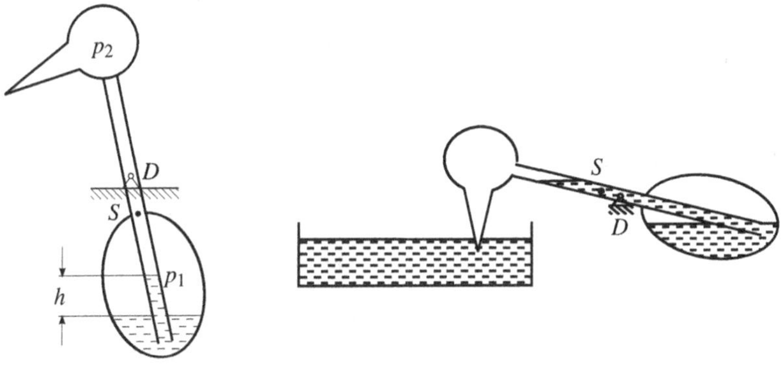

Fig. 5.17. Drinking bird: schematic arrangement with D–pivot point, S–center of gravity and arrangement before the drinking bird sits up.

that now the lower end of the riser pipe is released (Fig. 5.17). The liquid quickly flows back from the head to the body, the center of gravity S moves down, the drinking bird sits up again and swings back and forth for some time. These oscillations stimulate the evaporation at the head and the heat transfer from the environment to the body. Slowly, the liquid level in the riser pipe increases, the drinking bird tilts, and the game starts again.

5.4.3 Woodpecker Toy

An interesting example of a nonlinear oscillator with self-excitation is the woodpecker toy, which represents a nonsmooth dynamic system [53]. The toy consists of a rod, on which the swinging woodpecker slides down. In particular, it consists of a sleeve, which slides on the rod with clearance, and of the woodpecker, which is connected to the sleeve by a spring (Fig. 5.18a). The most important part of the toy is the sleeve with clearance, which induces a self-locking of the sleeve at specific tilting angles $\pm\vartheta_{k1}$. The kinetic energy of the downward movement is built-up from gravity g due to sliding along the rod; at the lower self-locking position $\vartheta = +\vartheta_{k1}$ (Fig. 5.18a), it is converted into vibrational energy for the woodpecker after subtraction of shock losses, because the z-degree of freedom of the downward movement is suddenly stopped. The woodpecker is now swinging to a maximum deflection and back. Reaching the tilting angle $+\vartheta_{k1}$ again, the self-locking and therefore the z-degree of freedom is released. Under the influence of gravity, slipping occurs but the woodpecker swings upwards $\left(\dot{\vartheta} < 0\right)$. At $\vartheta = -\vartheta_{k1}$, self-locking blocks the z-degree of freedom. For a specific $\vartheta = -\vartheta_{k2}\,(|\,\vartheta_{k2}\,| > |\,\vartheta_{k1}\,|)$, the woodpecker is forced into a rapid turnaround due to the pecker impact. Thus, at $\vartheta = -\vartheta_{k1}$, it is released from self-locking and moves down. Now, it swings downwards $\left(\dot{\vartheta} > 0\right)$ and enters self-locking at $\vartheta = +\vartheta_{k1}$. The cycle then starts again.

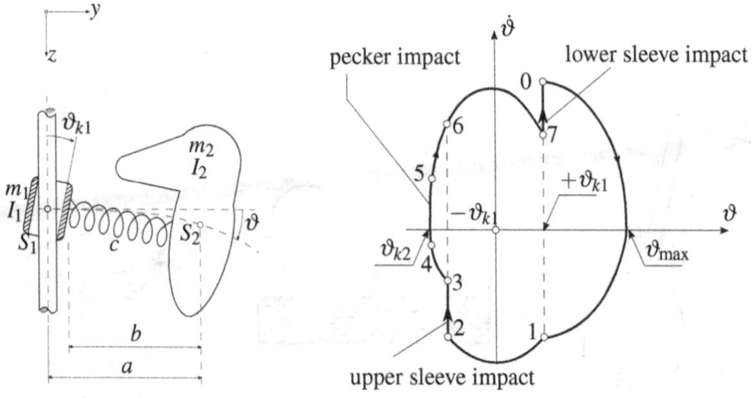

(a) schematic arrangement (b) schematic limit cycle

Fig. 5.18. Woodpecker toy.

The described motion is a nonlinear, self-excited vibration of storage-oscillator type with a stable limit cycle. The main switch is the sleeve with clearance, which controls the transformation of translational energy into vibrational energy as well as the build-up of translational energy from gravitation via self-locking. The pecker of the woodpecker is a further switch, which induces a fast reversion of the oscillation by the pecker impact. However, the pecker impact is not necessary for the functionality of the toy. The self-excitation mechanism is illustrated in Fig. 5.19.

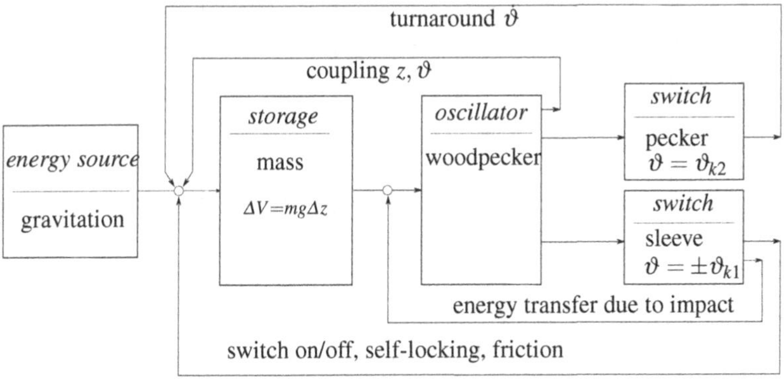

Fig. 5.19. Self-excitation mechanism.

The oscillation consists of five phases, which are connected with or without impacting behavior due to constraints. The phases themselves are described by differential equations for one or two degrees of freedom. For the connections with impacting self-locking, it is $\dot{z} = 0$ (Fig. 5.18a). The five phases of motion can be summarized as shown in Table 5.2 (Fig. 5.18b). A solution of the equations of motion connecting the different phases and using the periodicity results in the three-dimensional limit cycles according to Fig. 5.20.

The woodpecker toy is defined by $b = 0{,}015\,\text{m}$, $a = 0{,}025\,\text{m}$, $m_1 = 0{,}0003\,\text{kg}$, $m_2 = 0{,}0045\,\text{kg}$, $I_1 = 5 \cdot 10^{-9}\,\text{kgm}^2$, $I_2 = 7 \cdot 10^{-7}\,\text{kgm}^2$ as well as $c = 0{,}0056\,\text{Nm}$. The energy absorbed within the lower sleeve impact is about 79% of the supplied energy. The difference between theory and measurement of the limit-cycle frequency and of the falling height is only a few percent [53].

5.4.4 *Friction Oscillator*

Oscillation systems with dry friction (COULOMB friction) can be found in many technical applications, whereby the friction is often the source of the self-excited oscillations. The formation principle is always the same, such that we can discuss the principle with the example of the friction pendulum (FROUDE pendulum).

The friction pendulum consists of an engine, where the shaft rotates with constant angular velocity $\dot{\varphi}_w$ (Fig. 5.21). A pendulum is mounted on the shaft such that it can

Table 5.2. Phases of motion.

Fig. 5.18b	degree of freedom	$(\vartheta,\dot\vartheta)$ – begin	$(\vartheta,\dot\vartheta)$ – end	translation z
0 - 1	ϑ	$\vartheta=+\vartheta_{k1}$ $\dot\vartheta>0$	$\vartheta=+\vartheta_{k1}$ $\dot\vartheta<0$	$\dot z=0$
1 - 2	ϑ,z	$\vartheta=+\vartheta_{k1}$ $\dot\vartheta<0$	$\vartheta=-\vartheta_{k1}$ $\dot\vartheta<0$	$\dot z>0$
3 - 4	ϑ	$\vartheta=-\vartheta_{k1}$ $\dot\vartheta<0$	$\vartheta=-\vartheta_{k2}$ $\dot\vartheta<0$	$\dot z=0$
5 - 6	ϑ	$\vartheta=-\vartheta_{k2}$ $\dot\vartheta>0$	$\vartheta=-\vartheta_{k1}$ $\dot\vartheta>0$	$\dot z=0$
6 - 7	ϑ,z	$\vartheta=-\vartheta_{k1}$ $\dot\vartheta>0$	$\vartheta=+\vartheta_{k1}$ $\dot\vartheta>0$	$\dot z>0$

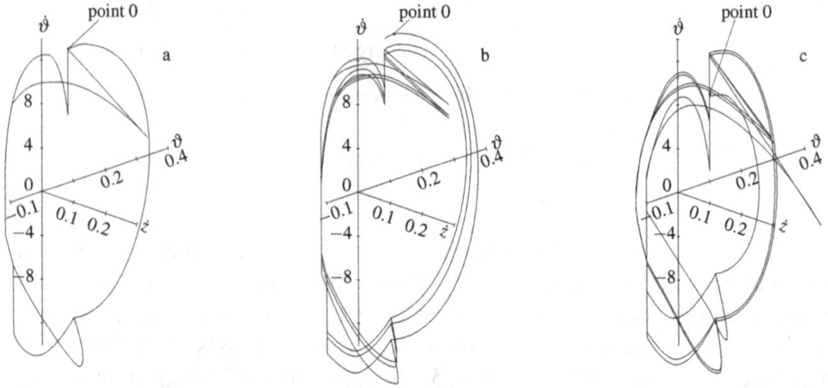

Fig. 5.20. Results: a reference limit cycle, b approaching from outside, c approaching from inside [53].

rotate freely but it is taken due to the friction forces in-between the pendulum sleeve and the shaft. This effect is limited by the point, at which the pendulum torque can no longer compensate the gravitational torque due to sticking forces.

The friction pendulum is thus performing the following motion patterns. For small sticking forces, the shaft will drive the pendulum up to a reversion point, where an equilibrium of the sticking forces, of the velocity-dependent friction forces at the shaft, and of the gravitational forces breaks down. Then, the pendulum moves against the shaft rotation until another reversion appears, where the force balance again allows a driving effect. The other extreme, namely a large friction force, results in a pendulum rotation with the shaft. Starting from these two basic motion patterns, an intermediate behavior is also possible which we do not discuss here [33, 42].

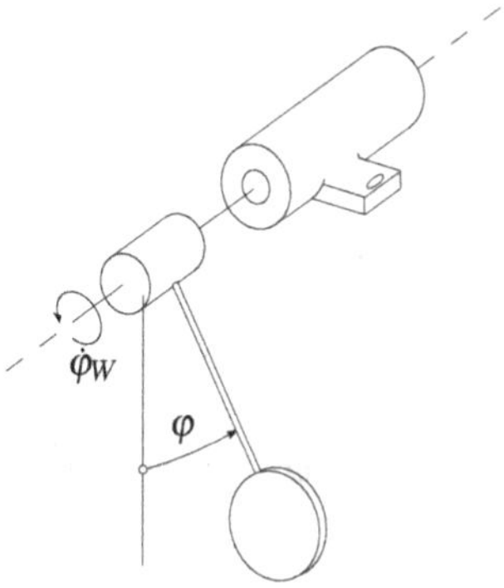

Fig. 5.21. Friction pendulum.

In the following, we consider the motion of the pendulum with low sticking friction. The characteristics of the dry friction decrease (Fig. 5.22). This allows self-excited oscillations to develop, because around the point of constant shaft velocity $\dot{\varphi}_w$, there is a larger friction torque for relative angular velocities $0 < \dot{\varphi}_r < \dot{\varphi}_w$ than for relative angular velocities $\dot{\varphi}_r > \dot{\varphi}_w$. This results in asymmetric oscillations. If the pendulum in Fig. 5.21 moves in the rotation direction of the shaft with $\dot{\varphi} > 0$, then the relative rotation satisfies $\dot{\varphi}_r < \dot{\varphi}_w$ because of $\dot{\varphi}_r = \dot{\varphi}_w - \dot{\varphi}$. According to Fig. 5.22, a larger friction torque is introduced than for the case $\dot{\varphi} < 0$ and $\dot{\varphi}_r > \dot{\varphi}_w$, that is energy is supplied to the pendulum by the friction torque.

The equation of motion for the friction pendulum in Fig. 5.21 is given by

$$J\ddot{\varphi} + d\dot{\varphi} + mgs\sin\varphi = R \tag{5.30}$$

with the following parameters: J pendulum moment of inertia, d velocity proportional damping, m pendulum mass, g gravitational constant, s distance of the center of gravity of the pendulum to its suspension point, R friction torque, and φ angle. The friction torque depends on the relative velocity $\dot{\varphi}_r$:

$$R = R(\dot{\varphi}_r) = R(\dot{\varphi}_w - \dot{\varphi}) \ . \tag{5.31}$$

With the dimensionless quantities

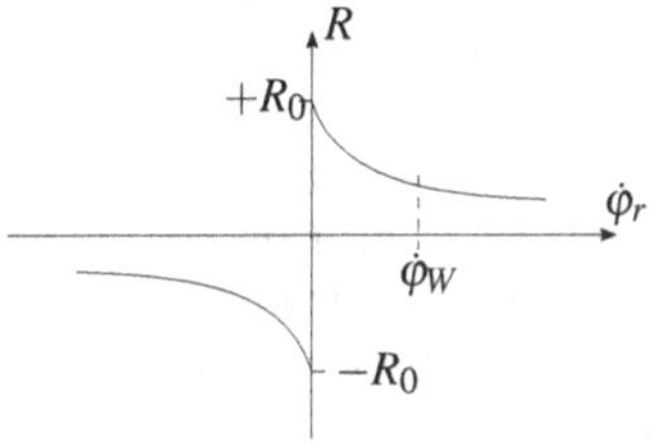

Fig. 5.22. Characteristics of the dry friction.

$$\frac{d}{J} = 2\delta\,, \quad \frac{mgs}{J} = \omega_0^2\,, \quad \frac{R}{J} = r\,, \tag{5.32}$$

one obtains

$$\ddot{\varphi} + 2\delta\dot{\varphi} + \omega_0^2 \sin\varphi = r\,(\dot{\varphi}_w - \dot{\varphi})\,, \tag{5.33}$$

which yields, by inserting the velocity and eliminating the time,

$$\dot{\varphi}\frac{\mathrm{d}\dot{\varphi}}{\mathrm{d}\varphi} = r\,(\dot{\varphi}_w - \dot{\varphi}) - 2\delta\dot{\varphi} - \omega_0^2 \sin\varphi\,. \tag{5.34}$$

With a known friction function r, the phase portrait can be derived. First, we recognize an equilibrium position with $\dot{\varphi} = 0$

$$r\,(\dot{\varphi}_w) - \omega_0^2 \sin\varphi = 0\,, \tag{5.35}$$

hence

$$\sin\varphi_0 = \frac{r\,(\dot{\varphi}_w)}{\omega_0^2} = \frac{R\,(\dot{\varphi}_w)}{mgs}\,. \tag{5.36}$$

Physically, this means that the pendulum adjusts to a value $\varphi_0 = \varphi_w > 0$ due to the friction between the pendulum sleeve and the motor shaft; this corresponds to a balance between friction torque R and gravitational torque (mgs).

If at larger pendulum amplitudes, the pendulum velocity equals the velocity of the shaft, that is $\dot{\varphi} = \dot{\varphi}_w$ and $\dot{\varphi}_r = 0$, this corresponds to a state of motion with maximum friction $R_0 = R(0)$ or $r_0 = r(0)$. This state of static friction or stiction may be left, if the damping and restoring forces just reach the value r_0. Then, the pendulum is released again from the shaft and starts to oscillate downwards. Such a *breakpoint* can be calculated using

$$r_0 - 2\delta\,\dot{\varphi}_w - \omega_0^2 \sin\varphi_1 = 0\,, \tag{5.37}$$

which results in

$$\sin\varphi_1 = \frac{r_0 - 2\delta\,\dot{\varphi}_w}{\omega_0^2}\,. \tag{5.38}$$

In the phase portrait, point φ_1 is on the *jump line* $\dot{\varphi}_r = 0$. For this value, the static friction jumps from $-R_0$ to $+R_0$ (Fig. 5.22). Each motion in the vicinity of the jump line leads into it (Fig. 5.23). If the pendulum velocity $\dot{\varphi}$ is smaller than $\dot{\varphi}_w$ and $\dot{\varphi}$ is increasing, the difference $\dot{\varphi}_w - \dot{\varphi}$ will decrease, the friction torque will increase (Fig. 5.22), and the motion will tend to $\dot{\varphi} = \dot{\varphi}_w$. However, an overshoot with $\dot{\varphi} > \dot{\varphi}_w$ results in a negative friction torque opposite the direction of shaft rotation, which, first, cannot be compensated by the gravitational torque, and which, second, ensures that the motion stays on the jump line with $\dot{\varphi} = \dot{\varphi}_w$.

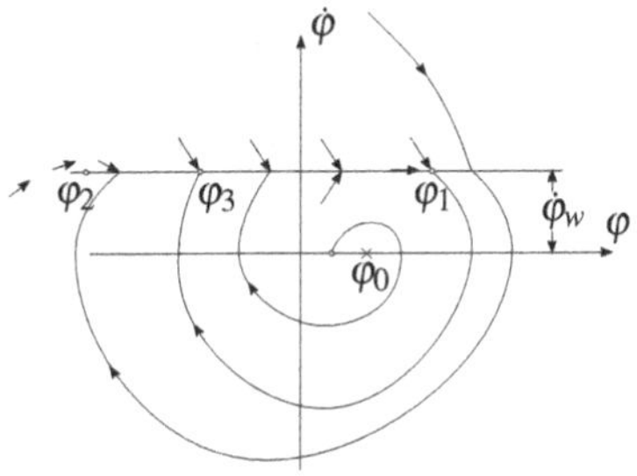

Fig. 5.23. Phase portrait for a friction pendulum.

The leftmost boundary of the jump line is a point satisfying $\dot{\varphi} = \dot{\varphi}_w$, $r = -r_0$, hence the corresponding (negative) angle is given by

$$\sin \varphi_2 = -\left(\frac{r_0 + 2\delta\dot{\varphi}_w}{\omega_0^2}\right) . \tag{5.39}$$

All phase curves, which lead into the line φ_2–φ_1, go to point φ_1 and spiral about the equilibrium position defined by (5.39). With high internal damping, the phase curves shrink to zero, with moderate internal damping, a limit cycle of the trajectory φ_1–φ_3 in Fig. 5.23 is created [33, 42].

5.4.5 *Kármán Vortex Street*

The periodic separation of vortices for round structures (wires, braces and the like), which are vertically passed by a fluid, can lead to resonance with the eigenfrequencies of the structure and to fractures. The most famous example is the collapse of the Tacoma Narrows Bridge on 7.11.1940 [32]. Less spectacular examples are vibrations of power transmission lines or the 'singing' of telephone lines [60]. In all cases, it is a self-excited oscillation, which takes its energy in cycles of vortex separations from the flow.

Theodore von KÁRMÁN was the first to study these periodic vortex structures (1911 in Göttingen, as an assistant of Ludwig PRANDTL). He concluded that only certain staggered vortex configurations are stable (Fig 5.24). Such a stable configuration has the width-spacing ratio

$$\frac{h}{l} = 0,283 . \tag{5.40}$$

We consider a flow around a cylinder with diameter D. The velocity, for which the stable vortex formation and separation occur, is equal to the flow velocity V_∞ minus the so-called induced velocity u. This is caused by the circulation Γ $\left[\mathrm{m}^2/\mathrm{s}\right]$ of the vortices:

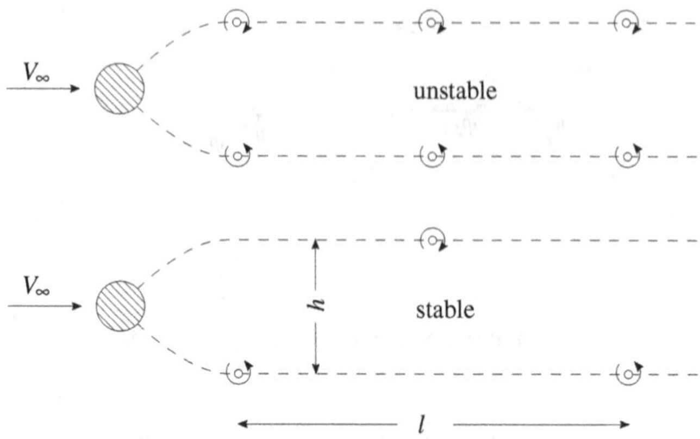

Fig. 5.24. Vortices in a flow following a circular cylinder.

$$V_w = V_\infty - u = V_\infty - \frac{1}{\sqrt{8}} \frac{\Gamma}{l} \,. \tag{5.41}$$

A new vortex pair is created after the time

$$T = \left(\frac{l}{V_\infty - u} \right) \,. \tag{5.42}$$

The staggered separation of vortices with frequency f may stimulate the body to a destructive oscillation, if its eigenfrequencies are excited. Therefore, we are interested in estimating the frequency f for a stable vortex street.

A regular vortex street occurs only for REYNOLDS numbers $Re = \frac{V_\infty D}{\nu}$ of about 60 to 5000 with the kinematic viscosity ν. For smaller REYNOLDS numbers, the flow is laminar, for larger REYNOLDS numbers, the flow is completely turbulent. In the given range of REYNOLDS numbers, there is a clear dependence of the REYNOLDS number on the dimensionless STROUHAL number

$$S = \frac{fD}{V_\infty} \tag{5.43}$$

due to measurements [60]. This relationship is shown in Fig. 5.25 as a summarizing result of measurements and calculations. For larger REYNOLDS numbers, there is a constant STROUHAL number $S = 0,21$.

For small cylinder diameters and moderate velocities, we evaluate frequencies in the acoustic range. The well-known 'singing' of telephone lines can be explained by this. With an air velocity $V_\infty = 10\,\mathrm{m/s}$ and a wire diameter $D = 2\,\mathrm{mm}$, the frequency satisfies

$$f = 0,21 \frac{10\,\mathrm{m/s}}{0,002\,\mathrm{m}} = 1050\,\mathrm{s}^{-1} \,. \tag{5.44}$$

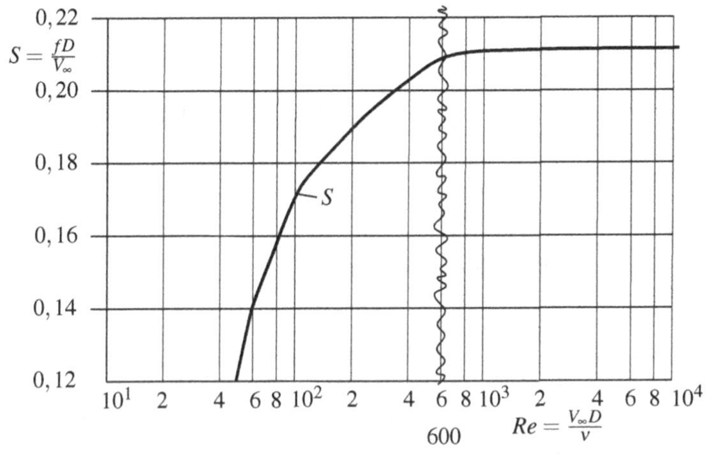

Fig. 5.25. STROUHAL number depending on the REYNOLDS number for the flow around circular cylinders [60].

Thereby, the REYNOLDS number is $Re \approx 1200$.

Theodore von KÁRMÁN illustrated the discovery of the vortex street and the work associated with the collapse of the Tacoma Narrows Bridge in his book *Die Wirbelstraße* in a very illustrative way [32].

5.4.6 Flutter

Flutter is known from aircraft, helicopters, and turbines. Like the KÁRMÁN vortex street, the phenomenon belongs to the flow-induced self-excitation mechanisms [8]. Flutter is a coupling effect between the flow conditions around the wing and the elastic properties of the wing; it is an aeroelastic problem that needs to be considered in any design of wings or blades. For this, very extensive and complicated calculations are necessary, which we can only indicate. We discuss the basic self-excitation mechanism with a conceptual model.

We model an airplane wing as a beam, which can perform bending and torsional vibrations, and we take into account only the first eigenfrequency and the associated first eigenmode of these bending and torsional vibrations (Chapter 3 and Fig. 5.26). The bending vibration ensures that the wing moves up and down, the torsional vibration generates a rotation of the wing cross section with respect to the inflow direction in each motion phase. Whether the wing gets to flutter or not depends on the eigenfrequencies of bending and torsion, as well as their relative phase. In Fig. 5.26, a case is shown in which bending and torsional vibration have the same eigenfrequency but a 90° shifted phase. When the wing oscillates upwards, it is rotated at the same time in such a way that a positive angle of attack and thus an additional wing lift is produced. When the wing oscillates downwards, it is rotated such that a negative angle of attack and thus a negative lift is generated. In such conditions, the

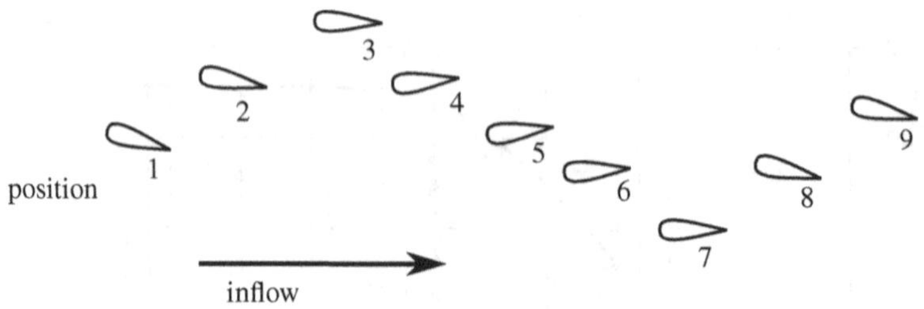

Fig. 5.26. Principle of flutter [8].

elastic vibrations of the wing are excited by the aerodynamics; unstable conditions may arise.

The self-excitation mechanism of flutter can be characterized as follows. Energy is taken from the flow energy of the surrounding fluid at the frequency of the elastic vibrations of the wing. This energy serves to maintain the oscillation process. The torsional vibrations act as a switch, setting the angle of attack of the wing at the correct moment to support the vibrations with additional positive or negative lift. The process depends on flight velocity. There are several critical velocities at which different forms of flutter can occur; also there are even more complicated flutter phenomena than the ones described above. To reduce the risk to the aircraft at such velocities, flutter must be reduced with appropriate design. For this, there are some general approaches.

A classical approach consists in adapting mass and stiffness distributions in combination with the aerodynamics of the wing to shift the resonances either out of the operational region or to avoid them. Against the background of modern computer software results are astonishing. A decoupling of elastic modes is difficult and the flutter problem not always completely solvable. A third important step is increasing damping. As we saw in Fig. 5.13 and in Fig. 5.14, a self-excited oscillation is generated by the balance of supplied and dissipated energy. The mean amplitude of the limit cycle will be smaller, the more energy is dissipated. In the limit, vibration no longer arises. As a general step in reducing oscillations, we require as much energy dissipation in the system, that is the wing, as possible. This idea, of course, also has constructive and function-related limits.

5.4.7 Pendulum Clock

The pendulum clock with GRAHAM escapement offers a classic example of self-excited oscillation. The energy source is a torsional spring or a coiled weight, the oscillator is a pendulum, and the switch is the anchor with its input and output pallets together with the escape wheel (Fig. 5.27). The escape wheel is held within a moment tension either by a spring or by a weight, such that it always tends to continue rotating in the sense of this external torque.

The GRAHAM escapement consists of an escape wheel with saw-shaped teeth, which is driven by the clock spring or by a weight, and an anchor with a fixed

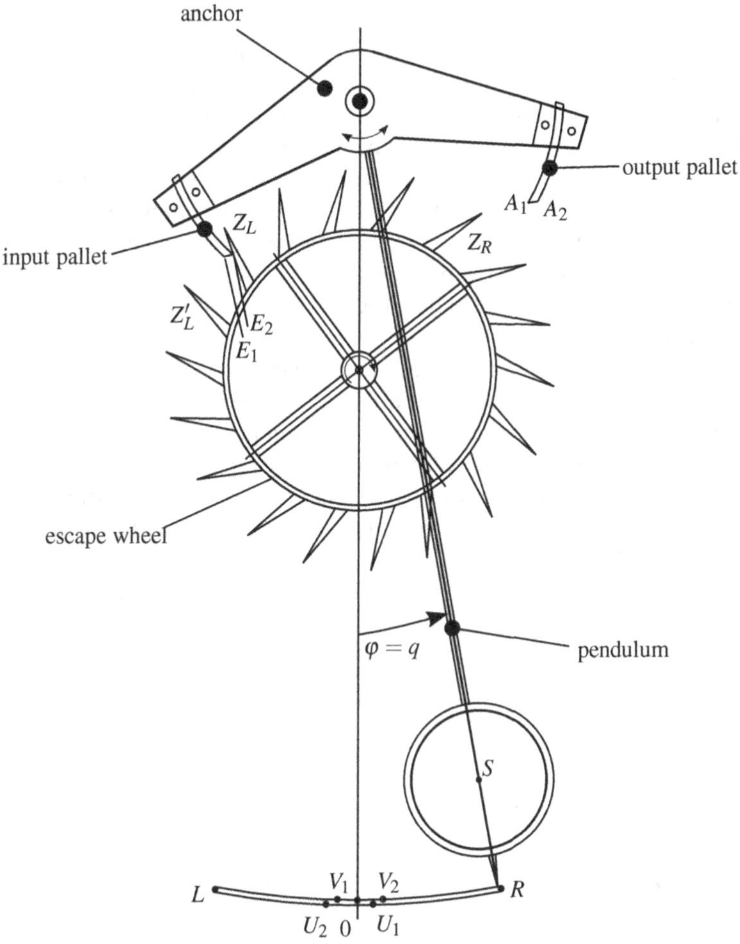

Fig. 5.27. GRAHAM escapement of a pendulum clock [33].

connection to the pendulum. Its two pallets, the input and the output pallet, comprise parts of a ring which is concentric to the pendulum axis [7, 19, 33].

Suppose now that the pendulum is about to swing from its outermost position R to the left; then, first, the tip Z_L of a tooth of the escape wheel slides on the cylindrical portion of the outer edge of the input pallet. Thereby, the escape wheel remains at rest and does not exert any force on the pendulum, if we disregard friction forces. When the pendulum reaches position U_1, which is on the right of the center 0, the tooth tip Z_L of the escape wheel changes from the outer edge of the input pallet to the transverse face E_1E_2 of this pallet. Now, the escape wheel can rotate further and exert a force on the pendulum, which results in a torque M_L acting on the pendulum in the sense of its motion direction. This phase ends when the tooth tip Z_L reaches point E_2, which corresponds to the pendulum position U_2 on the left of the center

0. At this moment, the escape wheel continues to rotate without constraint, until the tip Z_R of the tooth that is closest to the output pallet impacts the output pallet on its inner edge. A further rotation of the escape wheel is prevented. We assume, that the pendulum has not left position U_2, when the tooth tip Z_R hits the output pallet. When the pendulum swings to the left to its maximum position L, the tooth tip Z_R slides on the inner edge of the output pallet, whereby the escape wheel cannot rotate.

For the following half-cycle of the pendulum from position L to the right, the tooth tip Z_R slides on the inner edge of the output pallet (beyond position U_2, which has been reached by Z_R with the previous half-cycle), until it reaches point A_1 on the transverse face of the output pallet. The pendulum position achieved, V_1, is still left of the center 0. Now, the tooth tip Z_R slides on the transverse face A_1A_2 of the output pallet. Thereby, the escape wheel rotates and exerts a force on the pendulum, which results in a torque M_L acting in the direction of motion. When the tooth tip Z_R reaches point A_2, which corresponds to a pendulum position V_2 on the right of U_1, the escape wheel rotates without restriction and very rapidly, until the next tooth tip Z'_L strikes the outer edge of the input pallet. Then, the escape wheel stops again. As long as the pendulum finishes its half-cycle to the right, the tooth tip Z'_L slides on the

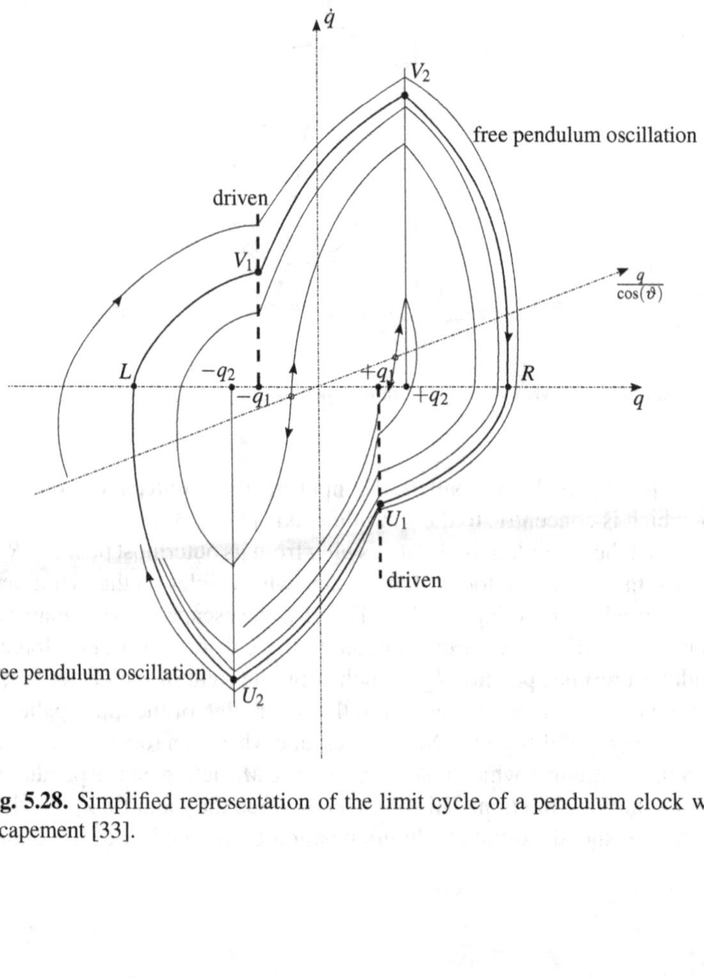

Fig. 5.28. Simplified representation of the limit cycle of a pendulum clock with GRAHAM escapement [33].

outer edge of the input pallet. When a position corresponding to the initial position R is reached, the whole procedure starts again with a new half-cycle to the left but the escape wheel rotated by one tooth. The input and the output pallet at the anchor are set in such a way, that the pendulum positions U_1 and V_1 as well as U_2 and V_2 are symmetric to the center 0, where $\overline{0U}_2 = \overline{0V}_2$ is larger than $\overline{0U}_1 = \overline{0V}_1$ [33].

Without going into detail on the extensive calculations connecting one phase to the other, let us consider the limit cycle, which is the basis for the stable pendulum motion. The angles q belonging to U_1, U_2, V_1, V_2 are measured from the perpendicular line and are denoted by $+q_1, -q_2, -q_1, +q_2$, with $q_2 > q_1 > 0$. Then in the limit cycle, the motion of the pendulum can be described as in the following section.

From position $V_2\,(+q_2)$, the pendulum swings to $U_1(+q_1)$ via $R\,(\dot{q} = 0)$ describing an elliptic arc. Along the arc from $(+q_1)$ to $(-q_2)$ corresponding to $(U_1 - U_2)$ it is driven. On the left side, the pendulum swings from $U_2\,(-q_2)$ to $V_1\,(-q_1)$ via $L\,(\dot{q} = 0)$ and is driven along the arc from $V_1\,(-q_1)$ to $V_2\,(+q_2)$. The connection of these phases shows a limit cycle, which is composed of elliptic spirals as a result of (dimensionless) dissipation ϑ on the pallet faces (Fig. 5.28 and [33]).

5.5 Parametrically Excited Vibrations

Parametrically excited vibrations arise from periodically time-varying parameters of the considered vibration system and influence machine dynamics only when the system is displaced from its undisturbed equilibrium position.

5.5.1 *Overview*

Typical practical examples of parametrically excited oscillations are asymmetries of rotors, offset within drive shafts or the periodically time-varying tooth stiffness in gear transmissions [8, 35, 61]. The latter example shows all typical characteristics of this vibration type. The tooth contact for a spur gear occurs along the so-called contact line, where the number of contacting teeth changes. This means that the tooth stiffness $k_v(t)$ fluctuates periodically about a mean value (Fig. 5.29). For a

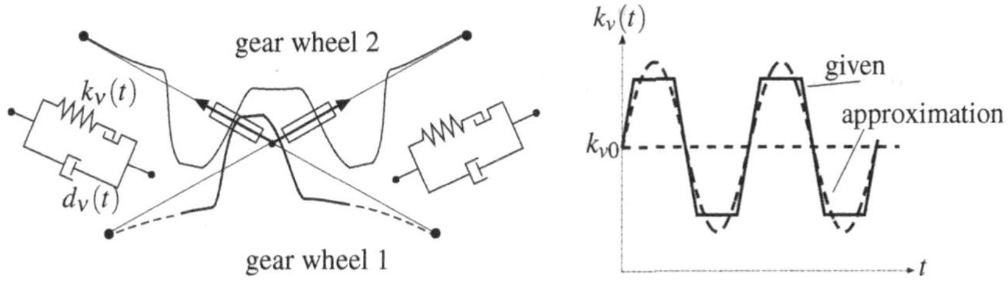

Fig. 5.29. Model of a single stage, spur gear transmission, and the corresponding tooth stiffness function with approximation.

mechanical model, the time-varying gear contact can be represented with corresponding springs and dampers (Fig. 5.29). Deriving the equations of motion, we obtain time-varying damping and stiffness matrices as typical characteristics of parametrically excited behavior.

Phenomena that can only be observed in parametrically excited systems and that are sometimes critical are *combination resonances*. For excitation of the system with angular frequencies Ω in the vicinity of specific combinations of eigen angular frequencies of the undamped time-invariant system, the amplitudes may increase. We obtain resonance phenomena and possible instabilities. These parameter and combination resonances occur for angular frequencies of the following structure:

$$\Omega = \frac{1}{p} (q_k \omega_k \pm q_l \omega_l) \ . \tag{5.45}$$

It is $(p, q_k, q_l) = 1, 2, 3, \ldots$ and $(k, l) = 1, 2, 3, \ldots f$ with the number of degrees of freedom f. The eigen angular frequencies of the undamped, time-invariant system are given by (ω_k, ω_l) with $d_v(t) = 0$ and $k_v(t) = k_{vo}$.

We have to avoid these exciting angular frequencies, which correspond to (5.45). It is not sufficient to analyze the vibrational behavior of a parametrically excited system only considering the eigen angular frequencies; this can lead to dangerous conclusions.

Sometimes parametrically excited vibrations are called *rheonomic vibrations*, a reference to rheonomic (time-variable) constraints in analytical mechanics. Depending on the type of system, we get *rheo-linear* or *rheo-nonlinear vibrations*.

5.5.2 Motion and Stability of Parametrically Excited Vibrations

5.5.2.1 Pendulum with Moving Suspension Point

If the suspension point A of a pendulum (Fig. 5.30 and Example 4.2) is moved up and down with the periodic time-variable acceleration $\ddot{a}(t)$, reaction forces $\ddot{a}(t)\mathrm{d}m$ will be created at the mass element of the pendulum. For such a physical pendulum, we get the equation of motion

$$J\ddot{\varphi} + ml\,(g + \ddot{a}) \sin\varphi = 0 \tag{5.46}$$

with moment of inertia J with respect to A and

$$\ddot{a} = \ddot{a}_0 \cos(\Omega_0 t) \ . \tag{5.47}$$

We linearize the equation of motion about the lower equilibrium point. With

$$\tau = \Omega_0 t \ , \quad \lambda = \frac{mlg}{\Omega_0^2 J} \ , \quad \gamma = \frac{ml\ddot{a}_0}{\Omega_0^2 J} \ , \quad \frac{\mathrm{d}}{\mathrm{d}\tau} = (\,\cdot\,)' \ , \tag{5.48}$$

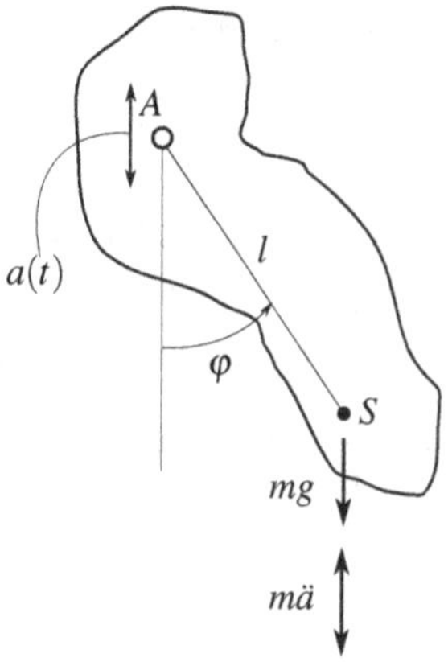

Fig. 5.30. Pendulum with driven suspension point.

we get the dimensionless form [33]

$$\varphi'' + (\lambda + \gamma \cos \tau)\, \varphi = 0\,. \tag{5.49}$$

5.5.2.2 Kane's Baby Shoes

A Stanford student observed, that a pair of small baby shoes hanging from the mirror of a car started to vibrate strongly in certain car situations and at certain frequencies. KANE used a model with two degrees of freedom to show that the rocking is a parametrically excited vibration [31]. Fig. 5.31 depicts the principal situation. The equations of motion for this model are

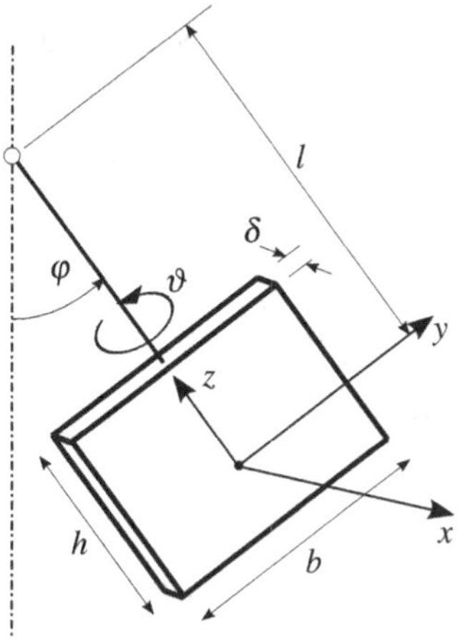

Fig. 5.31. Model for the baby shoes according to KANE [31].

$$\left(1+a\cos^2\vartheta\right)\ddot{\varphi}-a\dot{\vartheta}\dot{\varphi}\sin2\vartheta+(1+a)\Omega_0^2\sin\varphi=0\,, \tag{5.50}$$

$$\ddot{\vartheta}+\frac{\dot{\varphi}^2}{2}\sin2\vartheta=0\,. \tag{5.51}$$

Thereby, we assume that the shoe of mass m and moment of inertia $I_x \approx I_z, I_y \ll 1$ about its center of gravity is hanging from a massless string, such that $a=\frac{I_x}{ml^2}$ and $\Omega_0=\frac{mlg}{ml^2+I_x}$. To get a first approximation, we further assume that

- the minimal coordinate φ oscillates harmonically: $\varphi=\varphi_0\cos(\Omega_0 t)$,
- the minimal coordinate ϑ is small: $\vartheta \ll 1$.

From (5.51), we get

$$\ddot{\vartheta}+\left(\frac{\varphi_0^2\Omega_0^2}{2}\right)(1-\cos2\Omega_0 t)\,\vartheta=0\,. \tag{5.52}$$

After normalization with $\tau=2\Omega_0 t$ and $\frac{\mathrm{d}}{\mathrm{d}\tau}=(\,\cdot\,)'$, this yields

$$\vartheta''+(\lambda+\gamma\cos\tau)\,\vartheta=0 \tag{5.53}$$

with

$$\lambda=-\gamma=\frac{\varphi_0^2}{8}\,. \tag{5.54}$$

5.5.2.3 Mathematical Relationships

There are many more examples of parametrically excited vibrations [2, 10, 8, 26, 33, 35, 42, 44, 61]. The equations of motion of type (5.49) and (5.53) are typical. In the following, we discuss these ordinary, linear, and time-variant differential equations in detail.

For the case of an oscillator with one degree of freedom and parametric excitation, we generally get:

$$\ddot{x}+p_1(t)\dot{x}+p_2(t)x=0\,. \tag{5.55}$$

With the ansatz

$$x=ye^{-\frac{1}{2}\int p_1(t)\mathrm{d}t}\,, \tag{5.56}$$

equation (5.55) can be transformed into

$$\ddot{y}+P(t)y=0 \tag{5.57}$$

with $P(t)=p_2(t)-\frac{1}{2}\frac{\mathrm{d}}{\mathrm{d}t}[p_1(t)]-\frac{1}{4}p_1^2(t)$. As $p_1(t)$ and $p_2(t)$ are assumed to be periodic, also $P(t)$ is periodic:

$$P(t+T) = P(t)\,. \tag{5.58}$$

Equation (5.57) with (5.58) is called the HILL *differential equation*. For systems with multiple degrees of freedom, we obtain the same structure, however $y(t)$ is a vector and $P(t)$ a matrix.

A solution can be found with FLOQUET *theory* [61, 2]. In the case with one degree of freedom, we choose the ansatz

$$y(t) = C_1 e^{\mu_1 t} y_1(t) + C_2 e^{\mu_2 t} y_2(t)\,. \tag{5.59}$$

Thereby, y_1 and y_2 are periodic functions of time, C_1 and C_2 are constants, and μ_1 and μ_2 are so-called *characteristic exponents* of (5.57). These exponents depend on the values in (5.57), but not on the initial conditions. They define the stability behavior of the solution. If one of the characteristic exponents has a positive real part, the solution (5.59) increases without restriction; it is unstable. If the real parts of both exponents are negative, then y decreases to zero; the solution is (asymptotically) stable. In the limiting case, the real part of one exponent (or both exponents) may vanish. Then y is bounded without approaching zero asymptotically; y may be periodic in this case. For the study of vibrations, real exponents are of interest. Then, the regions of stable solutions are separated from the regions of unstable solutions by a limiting characteristic of pure periodic solutions. Thus, the search for unstable regions results in the determination of conditions for vanishing exponents, that is pure periodic solutions.

For some specific forms of the periodic functions $P(t)$, the solutions of (5.57) have been systematically analyzed:

$$P(t) = P_0 + \Delta P \cos(\Omega t)\;, \tag{5.60}$$

$$P(t) = P_0 + \Delta P \operatorname{sgn}(\cos(\Omega t))\;. \tag{5.61}$$

For the first case, the parameter fluctuates according to a harmonic law, for the second case, the changes are abrupt, such that $P(t)$ is a Meander function. With (5.60), the HILL differential equation transforms into a MATHIEU *differential equation*; with (5.61), it transforms into a MEISSNER *differential equation*.

Like in the above examples, we set

$$\tau = \Omega t\,, \quad \frac{\mathrm{d}}{\mathrm{d}\tau} = (\,\cdot\,)'\,, \quad \lambda = \frac{P_0}{\Omega^2}\,, \quad \gamma = \frac{\Delta P}{\Omega^2}\,. \tag{5.62}$$

This yields the normal form of the MATHIEU and the MEISSNER differential equation:

$$\text{MATHIEU differential equation} \quad y'' + (\lambda + \gamma \cos\tau)\,y = 0\,, \tag{5.63}$$

$$\text{MEISSNER differential equation} \quad y'' + (\lambda + \gamma \operatorname{sgn}(\cos\tau))\,y = 0\,. \tag{5.64}$$

The stability behavior of the MATHIEU differential equation has been analyzed by INCE and STRUTT [38]. The stability just depends on the parameters (λ, γ), whereby $\gamma = 0$ defines the periodic limiting case and $\gamma \neq 0$ leads to unstable regions, which grow for increasing γ (Fig. 5.32).

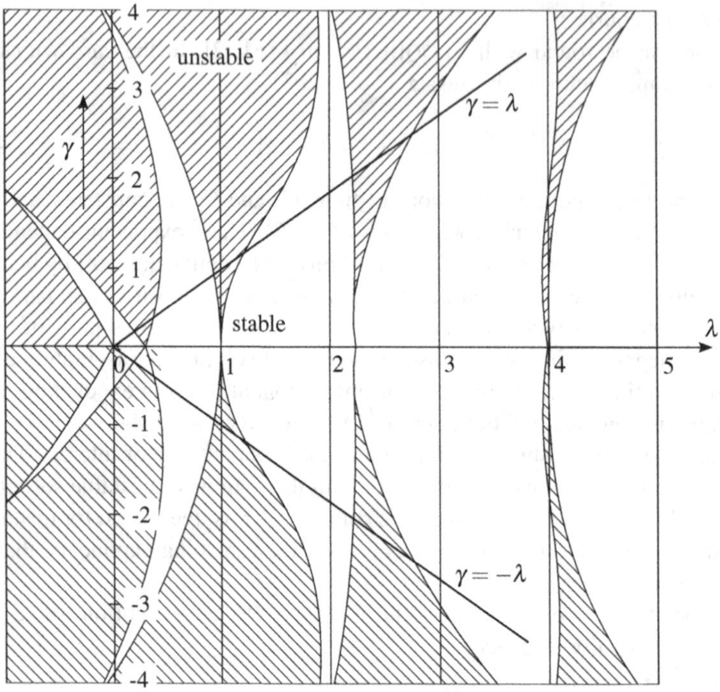

Fig. 5.32. Stability map according to INCE/STRUTT for the MATHIEU differential equation.

For negative λ and $\gamma = 0$, there are no parametric excitations and the system is unstable. If we consider an oscillator with constant, nonvanishing γ, its representation in the stability map will move along a parallel to the λ-axis for changing λ. For $\lambda > 0$, unstable regions may be crossed. In practice, the oscillator is stable for $\gamma = 0$ and may become unstable for $\gamma \neq 0$ and specific values of λ. The oscillating part with γ may have a stability reducing effect. In the case $\lambda < 0$, the oscillator is unstable for $\gamma = 0$ and may become stable for $\gamma \neq 0$. The oscillating part has a stabilizing effect.

The tips of the unstable regions touch the abscissa (λ-axis) at the values

$$\lambda = \left(\frac{n}{2}\right)^2 \quad (n = 1, 2, \ldots) \,. \tag{5.65}$$

The width of the regions – and therefore also their practical significance – reduces with increasing n. This can be explained by damping mechanisms, which have not been considered within the present analysis but which occur for real oscillators. They yield decreasing unstable regions [42].

In many cases, the area around the origin $\lambda = \gamma = 0$ of the stability map is of interest. Here, the boundary between the stable and unstable regions can be sufficiently approximated by functions $\lambda = \lambda(\gamma)$. For the first three boundary curves, we summarize without proof:

$$\lambda_1 \approx -\frac{1}{2}\gamma^2\,, \tag{5.66}$$

$$\lambda_2 \approx \frac{1}{4} - \frac{\gamma}{2}\,, \tag{5.67}$$

$$\lambda_3 \approx \frac{1}{4} + \frac{\gamma}{2}\,. \tag{5.68}$$

Fig. 5.33 shows an enlarged part of the stability map with the approximating boundary curves drawn as dotted lines.

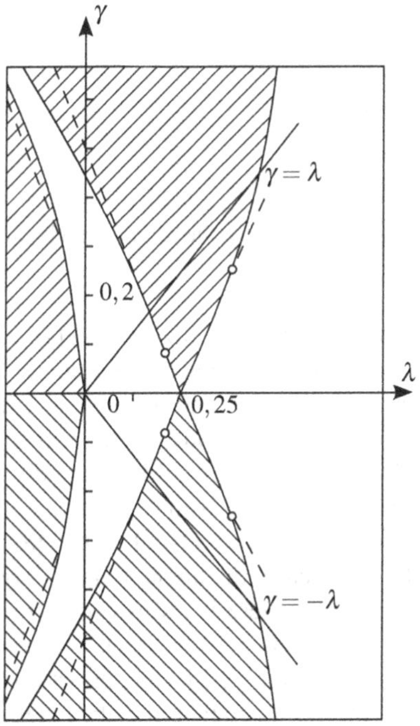

Fig. 5.33. Local stability map according to (5.66)-(5.68).

We return to the (linearized) examples. For the *pendulum with moving suspension point* (Section 5.5.2.1), we have to distinguish two cases, the hanging and the standing (inverted) pendulum. Equation (5.49) has been derived by linearization about the lower equilibrium point; for the hanging pendulum, we have $\lambda > 0$. If the value for λ satisfies $0 < \lambda < 0{,}25$, the oscillator will become unstable with increasing excitation amplitude γ (Fig. 5.32). Depending on the value of λ, the process will run at different speeds. If we keep γ constant and vary λ with the frequency ($\Omega = \Omega_0$), different stable and unstable regions are crossed along a horizontal line.

For the standing pendulum, the linearization has to be done about the upper equilibrium point: $\lambda < 0$. According to Figs. 5.32 and 5.33, only a small region of stable oscillations is possible. Without driving the suspension point ($\gamma = 0$), the pendulum is located in an unstable equilibrium point because of the center of gravity lying above the suspension point. It is remarkable that the unstable upper equilibrium position of the pendulum for a resting suspension point can be stabilized by appropriate vibration of the suspension point. This means that the pendulum may perform stable oscillations for small displacements from this equilibrium position.

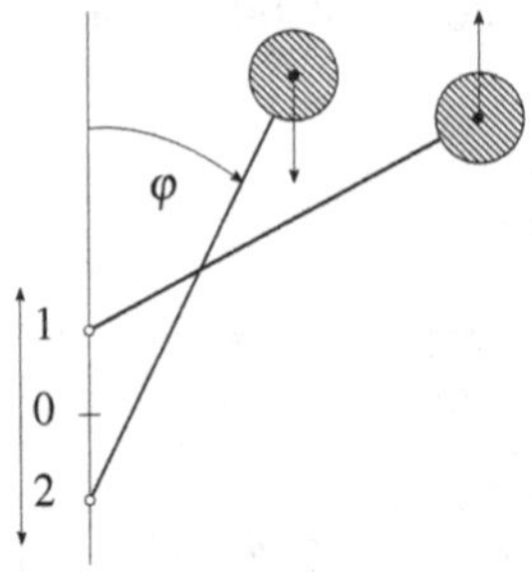

Fig. 5.34. Explanation of the stabilizing effect for the pendulum with driven suspension point.

We give a physical explanation for this stabilizing effect (Fig. 5.34). We consider a periodic motion between points 1 and 2, where the velocity vanishes in points 1 and 2. It is positive in an upward direction and negative in a downward direction. Then, the acceleration in the area $0-2-0$ is positive in an upward direction and the acceleration in the area $0-1-0$ is negative in a downward direction. As a reaction, there is an acceleration of the pendulum mass in a downward direction for the area $0-2-0$ of the suspension point; for the area $0-1-0$, there is an acceleration in an upward direction. For the latter case, the angle φ has a larger mean value than for the first case. As a consequence, there is always some rest acceleration in an upward direction and thus a torque left, which directs the pendulum to the upper equilibrium position. We can call this torque a jarring direction torque. If this torque is larger than the torque due to the gravitational force, then the pendulum will stay in the upper equilibrium positions and will not be destabilized by small perturbations.

The *jarring direction torque* may generate a shift of a beacon course when exciting in a corresponding direction. Thus, hanging measurement equipment might then be sensible in a direction of acceleration which is not desired.

For KANE*'s baby shoes* (Section 5.5.2.2), it is $\lambda = -\gamma \geq 0$. The relevant line $\gamma = -\lambda$ is shown in Figs. 5.32 and 5.33. With increasing amplitude φ_0, the double pendulum crosses different stable and unstable regions. If we transfer these regions into an amplitude stability diagram with φ_0 as the polar angle, the single regions can be highlighted; they can be easily tested experimentally (Fig. 5.35).

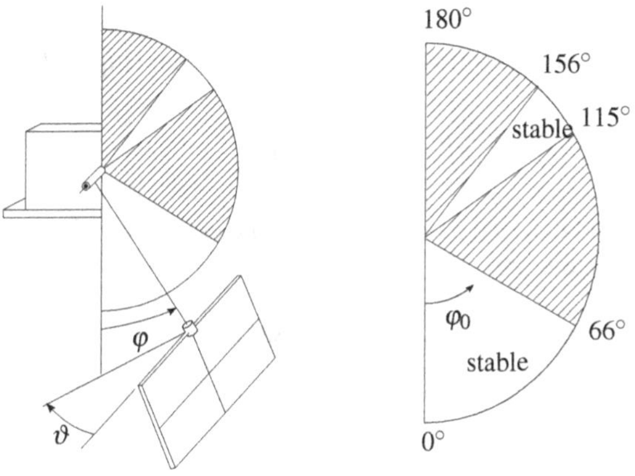

Fig. 5.35. KANE's baby shoes: typical stability regions for specific geometry and material values.

5.5.3 Examples

In the following, we consider classical cases for parametrically excited oscillations in practice.

5.5.3.1 Jeffcott Rotor with Stiffness Asymmetries

As an example of rotating machinery components with asymmetric behavior, we consider the so-called *Jeffcott rotor* [73] with nonsymmetric shaft stiffness. The rotor is assumed to be a point mass, the mass of the elastic shaft is neglected or added to the point mass. The shaft is mounted statically determinate and the speed Ω of the rotor is constant (Fig. 5.36). For the description, we use an inertial coordinate system I and a co-rotating coordinate system R with the same origin. If we consider

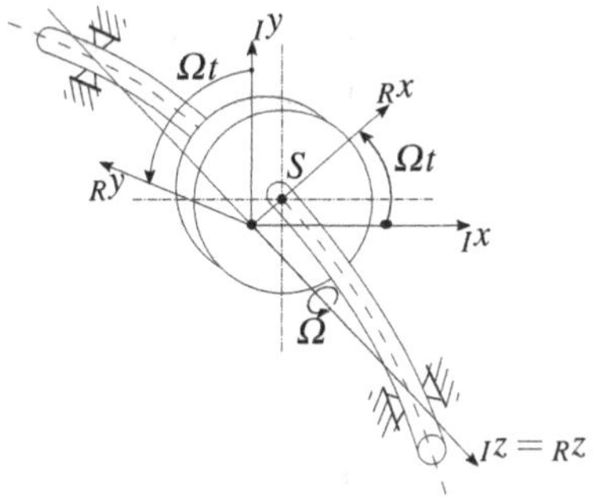

Fig. 5.36. Model of a Jeffcott rotor.

a stiffness asymmetry of the elastic shaft, the elastic restoring forces will act on the point mass of the rotor in the co-rotating coordinate system (Fig. 5.37):

$$ {}_R F_c = -\begin{pmatrix} c_x\, {}_R x \\ c_y\, {}_R y \end{pmatrix} . \tag{5.69} $$

Neglecting internal and external damping forces, the equations of motion in the co-rotating coordinate system are

$$ m_R\ddot{x} - 2m\Omega\, {}_R\dot{y} + \left(c_x - m\Omega^2\right) {}_R x = 0 , \tag{5.70} $$

$$ m_R\ddot{y} + 2m\Omega\, {}_R\dot{x} + \left(c_y - m\Omega^2\right) {}_R y = 0 . \tag{5.71} $$

With the abbreviations

$$ \varepsilon_S = \frac{c_x - c_y}{c_x + c_y} , \quad \omega_x^2 = \frac{c_x}{m} , \quad \omega_y^2 = \frac{c_y}{m} , \tag{5.72} $$

$$ \overline{\omega}^2 = \frac{\omega_x^2 + \omega_y^2}{2} , \quad \Omega_0 = \frac{\Omega}{\overline{\omega}} \tag{5.73} $$

and the transformation

$$ \tau = \overline{\omega} t , \quad \frac{\mathrm{d}}{\mathrm{d}t} = \overline{\omega} \frac{\mathrm{d}}{\mathrm{d}\tau} = \overline{\omega} (\cdot)' , \tag{5.74} $$

these equations can be written as

$$ {}_R\mathbf{z}'' + {}_R\mathbf{G}\, {}_R\mathbf{z}' + {}_R\mathbf{K}\, {}_R\mathbf{z} = \mathbf{0} \tag{5.75} $$

with

$$ {}_R\mathbf{z} = \begin{pmatrix} {}_R x \\ {}_R y \end{pmatrix} , \quad {}_R\mathbf{G} = \begin{pmatrix} 0 & -2\Omega_0 \\ 2\Omega_0 & 0 \end{pmatrix} , \quad {}_R\mathbf{K} = \begin{pmatrix} 1 + \varepsilon_S - \Omega_0^2 & 0 \\ 0 & 1 - \varepsilon_S - \Omega_0^2 \end{pmatrix} . \tag{5.76} $$

For practical reasons, the dynamic behavior of the Jeffcott rotor in Fig. 5.36 is analyzed in the inertial coordinate system, which makes sense for estimating the load

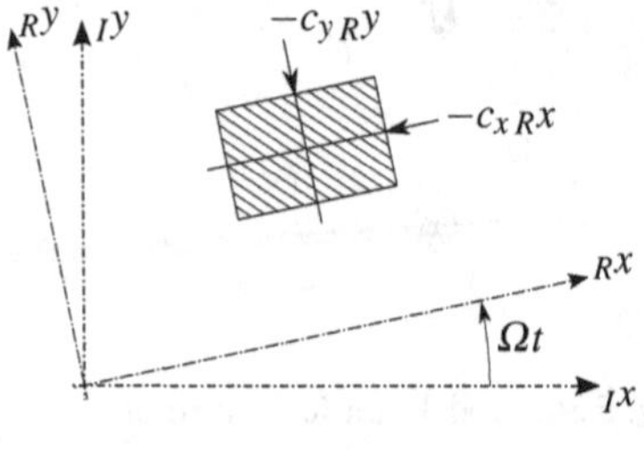

Fig. 5.37. Stiffness asymmetry and its effect on the rotor.

on the bearings. The transformation from the co-rotating to the inertial coordinate system is a rotation about the common z-axis:

$$ {}_R\mathbf{z} = \mathbf{A}_{RI}\,{}_I\mathbf{z} \quad \text{with} \quad \mathbf{A}_{RI} = \begin{pmatrix} \cos(\Omega_0\tau) & \sin(\Omega_0\tau) \\ -\sin(\Omega_0\tau) & \cos(\Omega_0\tau) \end{pmatrix} . \tag{5.77} $$

Inserting this transformation in (5.75), we obtain a linear system of differential equations with periodic coefficients:

$$ {}_I\mathbf{z}'' + {}_I\mathbf{K}\,{}_I\mathbf{z} = \mathbf{0} \quad \text{with} \quad {}_I\mathbf{K} = \begin{pmatrix} 1+\varepsilon_S\cos 2\Omega_0\tau & \varepsilon_S \sin 2\Omega_0\tau \\ \varepsilon_S\sin 2\Omega_0\tau & 1-\varepsilon_S\cos 2\Omega_0\tau \end{pmatrix} . \tag{5.78} $$

This is only a formally parametrically excited oscillation of a system with two degrees of freedom and, at the same time, the simplest possible representation of a rotating machine component with nonsymmetric stiffness. Equation (5.78) could be solved with the help of FLOQUET theory [61]. However in this case, a stability prediction can also be obtained from the equations of motion (5.75) in the co-rotating coordinate system with constant coefficients. The ansatz ${}_R\mathbf{z} = {}_R\mathbf{z}_0 \exp(\lambda\tau)$ yields the characteristic equation

$$ \lambda^4 + 2\left(1+\Omega_0^2\right)\lambda^2 + \left[\left(1-\Omega_0^2\right)^2 - \varepsilon_S^2\right] = 0 . \tag{5.79} $$

According to the STODOLA criterion (2.121), all the coefficients of the characteristic equation have to be positive. This is satisfied for the first two coefficients, (1) and $\left(2\left(1+\Omega_0^2\right)\right)$. For the last expression, we have to require

$$ \varepsilon_S^2 < \left(1-\Omega_0^2\right)^2 \quad \text{for stability} . \tag{5.80} $$

This result is sketched in Fig. 5.38. We recognize, that the system becomes unstable for the undamped, nonsymmetric Jeffcott rotor and $\Omega_0 = \frac{\Omega}{\varpi} \to 1$. However, the always present internal and external dampings in practice ensure that a stable operation is possible for $\Omega_0 \to 1$. Therefore, the asymmetries must not be too large. A case with damping is presented in Fig. 5.38 using a dotted line.

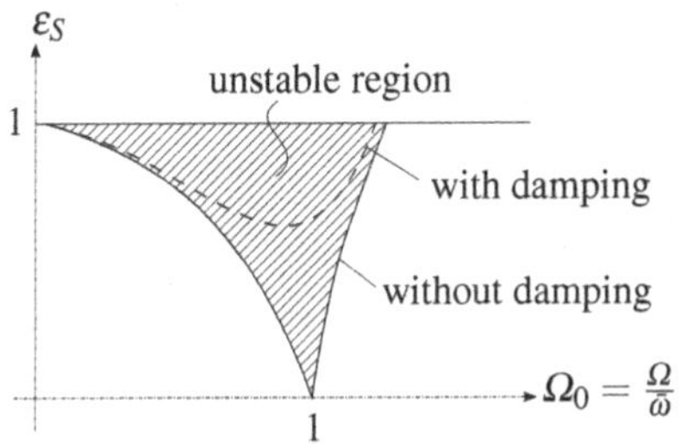

Fig. 5.38. Stability map for the Jeffcott rotor with stiffness asymmetry.

Despite the simple model assumptions, equation (5.78) and the stability map in Fig. 5.38 already show some essential features of vibrations of rotor systems. According to (5.78), the effective shaft stiffness varies twice during a period between its largest value $(1+\varepsilon_S)$ and its smallest value $(1-\varepsilon_S)$. This procedure results in a continuously changing load on the shaft and on the bearings, such that just because of this effect the asymmetries have to be minimized. In addition according to Fig. 5.38, the stability of the rotor system will be risky, if a *critical* speed (in the case $\Omega_0 = 1$) has to be passed. Even with existing dampings, instability peaks occur in such situations, which go very far downwards and therefore also require minimal asymmetries (see next section).

5.5.3.2 Stability and Control of Rotors

A more sophisticated example of a rotor with asymmetric stiffness than in the last section has been treated in [2]. Fig. 5.39 shows an ultracentrifuge model, where the rotor stiffness is unbalanced due to an elastic rectangular rod. The stability behavior of this rotor with 18 degrees of freedom has been analyzed and improved by means of some control. The control forces are generated by magnetic bearings, which are not shown. Without going into the very complex modeling and design of a controller, we show only the most important results.

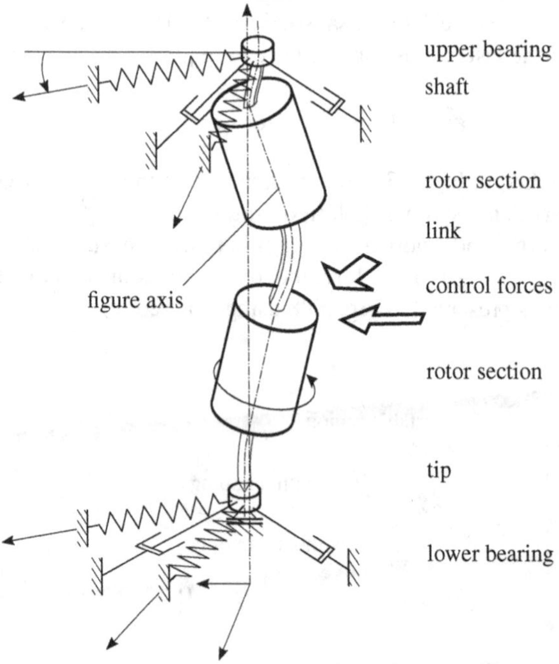

Fig. 5.39. Rotor with asymmetric stiffness [2].

The diagram on the left of Fig. 5.40 shows the stability behavior of the rotor without control. We can think of it as a kind of INCE-STRUTT diagram for a system with many degrees of freedom (analogous to Fig. 5.32). The abscissa indicates the speed, the ordinate gives the dimension ε_S (5.72) for the asymmetry of the rotor. One can see some single and combination resonances A-H, for which the motion becomes unstable for minor asymmetries (tip down at A). A stable operation of the rotor without controller is no longer possible from speeds $\omega \approx 210\,{}^1\!/\!_s \approx 2000\,\text{Upm}$. With an optimized control, the stability can be extended to $\omega \approx 1800\,{}^1\!/\!_s \approx 17000\,\text{Upm}$ (diagram to the right in Fig. 5.40). The diagram on the lower right in Fig. 5.40 is a close-up of the upper right diagram in the low speed range.

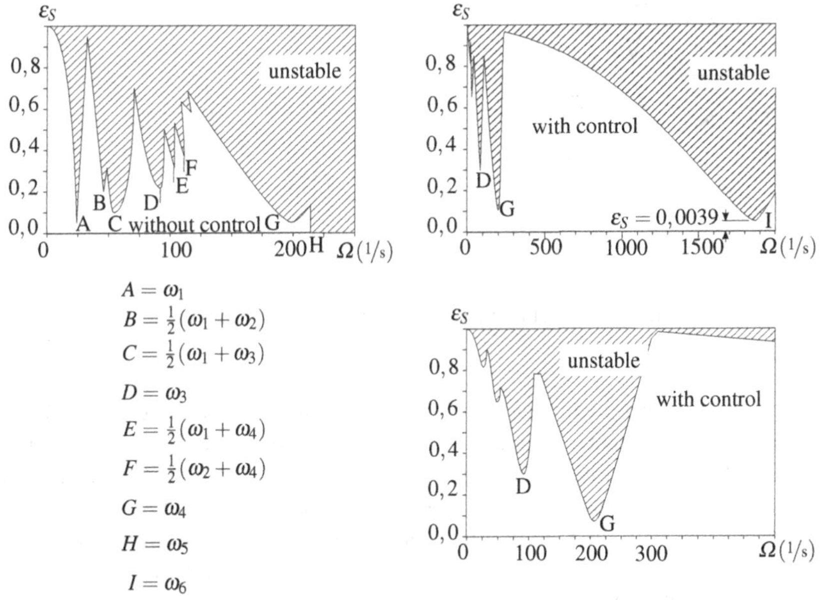

Fig. 5.40. Improving the stability of an asymmetric rotor with a control.

The results clearly show that vibrations of moving machine parts may pose a serious risk. Diagram 5.40 for the noncontrolled case proves that combination resonances (for $\frac{\omega_1+\omega_3}{2}$) can be just as dangerous as single resonances (for ω_1).

5.5.3.3 Gear Drive

In the overview 5.1, the periodically time-varying stiffness of a gear drive was presented as a typical example of a parametrically excited vibration. In the manual transmission of a car as a component of the entire drive train, the teeth contacts of the locked gears create parametric excitation in the entire system. Fig. 5.41 shows the mechanical model of such a drive train with 26 degrees of freedom and highlights the detailed conditions for a single tooth contact (a locked gear consists of

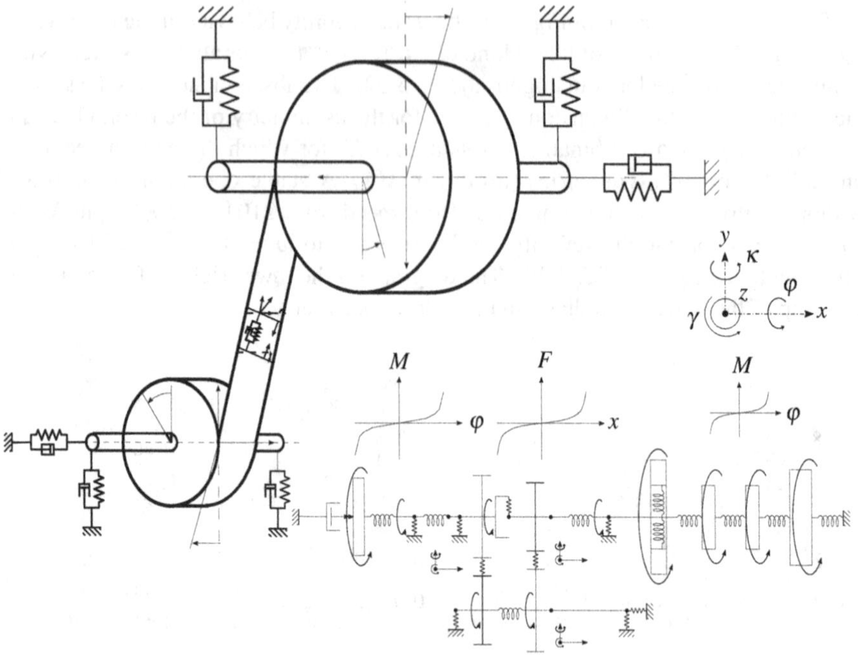

Fig. 5.41. Mechanical model of a passenger car drive train and a single gear ratio.

two meshings: *constant* and gear ratio). The problem of such a drive train is less a question of stability than a question of resonances, particularly because it is operated over a large speed range. The parametrically excited vibrations yield additional critical resonances because of the possible combination resonances. The study of these resonances with the help of a mechanical model and the resulting simulation model is not treated. Fig. 5.42 shows a typical result of a parametrically excited vibration (resulting from teeth contacts).

The time-varying tooth stiffness is the dotted curve. It shows about 10 downward tips during a full revolution. These stiffness peaks cause a shock to the teeth in contact and displace them abruptly. The structural damping ensures that the displacing impact almost completely decays until the next stiffness peak. Then the process begins again.

5.5.3.4 Pendulum with Elastic String

The pendulum with elastic string (Fig. 5.43) shows a type of coupled oscillation, which cannot be calculated from simple superpositions. The elastic suspension provides a kind of parametric excitation, which results in a periodic change of the stroke movement and swinging if the parameter values are adjusted appropriately. To understand the process, we derive the equations of motion. The kinetic and potential energy of the pendulum are

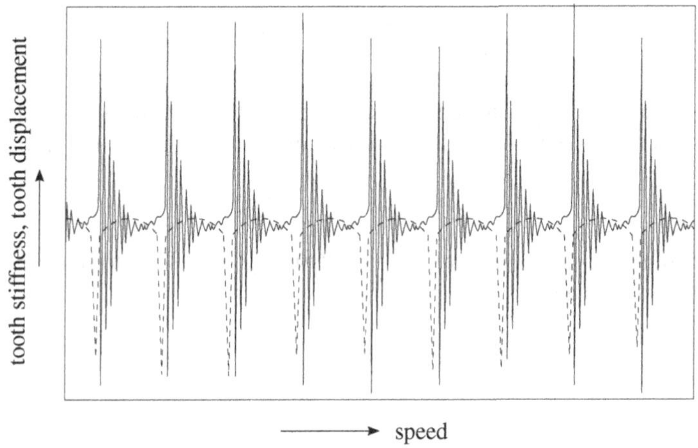

Fig. 5.42. Tooth stiffness (dotted) and tooth displacement (solid line) in a meshing during a revolution.

$$T = \frac{m}{2}\left[\dot{x}^2 + (R+x)^2\,\dot{\vartheta}^2\right], \tag{5.81}$$

$$V = mg\,(R+x)\,(1-\cos\vartheta) + \frac{1}{2}cx^2\,. \tag{5.82}$$

Applying LAGRANGE's equations of the second kind (1.178) for the generalized coordinates $\mathbf{q} = (x, \vartheta)^T$ and assuming small angles $\vartheta \ll 1$ and velocities $\dot{\vartheta} \ll 1$, we obtain the equations of motion in the linearized form

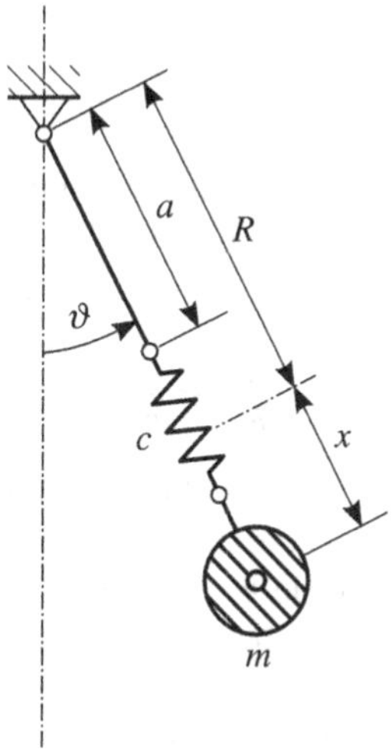

Fig. 5.43. Pendulum with elastic string.

$$\ddot{x} + \omega_x^2 x = 0\,, \tag{5.83}$$

$$\left[1 + \left(\frac{x}{R}\right)\right]\ddot{\vartheta} + 2\left(\frac{\dot{x}}{R}\right)\dot{\vartheta} + \omega_\vartheta^2 \vartheta = 0 \tag{5.84}$$

with $\omega_x^2 = \frac{c}{m}$ and $\omega_\vartheta^2 = \frac{g}{R}$. The length R corresponds to that of the static equilibrium position. As can be seen from (5.84), the harmonic oscillation of the x-coordinate

$$x = x_0 \cos(\omega_x t + \varphi) \tag{5.85}$$

acts as a parametric excitation for the ϑ-oscillation. If we restrict to small oscillation amplitudes with the ansatz

$$\vartheta \approx \vartheta_0 \cos(\omega_\vartheta t + \psi) + \Delta\vartheta\,, \tag{5.86}$$

we obtain the relationship

$$\begin{aligned}\Delta\ddot{\vartheta} + \omega_\vartheta^2 \Delta\vartheta \approx \frac{1}{2}\left(\frac{x_0}{R}\right)\vartheta_0\omega_\vartheta \{&(\omega_\vartheta - 2\omega_x)\cos((\omega_x + \omega_\vartheta)t + (\varphi - \psi)) \\ &+ (\omega_\vartheta + 2\omega_x)\cos((\omega_x - \omega_\vartheta)t + (\varphi + \psi))\}\end{aligned} \tag{5.87}$$

from (5.84) and (5.85) neglecting quadratic and higher terms in $\Delta\vartheta$ and $\frac{x_0}{R}$. This can be interpreted as an undamped, forced oscillation according to Section 5.3. Hence, it has the solution

$$\Delta\vartheta = V_1 \cos((\omega_x - \omega_\vartheta)t + (\varphi - \psi)) + V_2 \cos((\omega_x + \omega_\vartheta)t + (\varphi + \psi)) \tag{5.88}$$

with

$$V_1 = \frac{1}{2}\left(\frac{x_0}{R}\right)\vartheta_0\left(\frac{\omega_\vartheta}{\omega_x}\right)\left(\frac{\omega_\vartheta - 2\omega_x}{2\omega_\vartheta - \omega_x}\right), \tag{5.89}$$

$$V_2 = \frac{1}{2}\left(\frac{x_0}{R}\right)\vartheta_0\left(\frac{\omega_\vartheta}{\omega_x}\right)\left(\frac{\omega_\vartheta + 2\omega_x}{2\omega_\vartheta + \omega_x}\right). \tag{5.90}$$

It can be concluded that for small oscillation amplitudes, the stroke and swing oscillations are decoupled as long as $2\omega_\vartheta \neq \omega_x$. No vibration excites the other, and $\Delta\vartheta$ is very small because of $\frac{x_0}{R}\vartheta_0 \ll 1$. However, if $2\omega_\vartheta$ approaches ω_x, that is

$$4\left(\frac{g}{R}\right) \approx \frac{c}{m}, \tag{5.91}$$

the amplification function V_1 increases and the oscillation in the x-direction influences strongly the ϑ-oscillation. There is a periodic exchange between stroke and swing motion. These phenomena can be realized and observed with a tuned pendulum according to Fig. 5.43. It can be seen that for large oscillation amplitudes, there is always a coupling between stroke and swing oscillation. In (5.83)-(5.84), the nonlinear couplings between both motion directions have been neglected.

References

[1] Angeles, J.: Dynamic Response of Linear Mechanical Systems. Springer, New York (2012)

[2] Anton, E.: Stabilitätsverhalten und Regelung von parametererregten Rotorsystemen, Fortschritt-Berichte VDI: Reihe 8, Mess-, Steuerungs- u. Regelungstechnik, vol 67. VDI Verlag, Düsseldorf (1984)

[3] Arnold, V.: Mathematical Methods of Classical Mechanics, 2nd edn. Springer, Berlin (1997)

[4] Bathe, K.J.: Finite element procedures. MIT, Cambridge (2007)

[5] Bauchau, O.: Flexible Multibody Dynamics. Springer, Berlin (2010)

[6] Becker, E., Bürger, W.: Kontinuumsmechanik. Teubner, Stuttgart (1975)

[7] Berthoud, F.: Anweisung zur Kenntnis, zum Gebrauch und zur guten Haltung der Wand- und Taschenuhren. Reprint, Zentralantiquariat der DDR, Meissen (1818)

[8] Bishop, R.: Schwingungen in Natur und Technik. Teubner, Stuttgart (1985)

[9] Braess, D.: Finite elements: theory, fast solvers, and applications in elasticity theory, 3rd edn. Cambridge University Press, Cambridge (2007)

[10] Bremer, H.: Dynamik und Regelung mechanischer Systeme. Teubner, Stuttgart (1988)

[11] Bremer, H.: Elastic Multibody Dynamics. Springer, New York (2008)

[12] Brenner, S., Scott, R.: The mathematical theory of finite element methods. Springer, Berlin (1994)

[13] Courant, R.: Variational methods for the solution of problems of equilibrium and vibrations. Bull. Amer. Math. Soc. 49:1–49:23 (1943)

[14] Courant, R., Hilbert, D.: Methods of mathematical physics. Wiley, New York (1989)

[15] Dresig, H., Holzweißig, F.: Dynamics of machinery: theory and applications. Springer, Berlin (2010)

[16] Fischer, U., Stephan, W.: Prinzipien und Methoden der Dynamik. VEB Deutscher Verlag der Wissenschaften, Leipzig (1972)

[17] Galerkin, B.: Rods and plates. series occurring in various questions concerning the elastic equilibrium of rods and plates. Engineers Bulletin (Vestnik Inzhenerov) 19:897–19:908 (1915)

[18] Gander, M., Wanner, G.: From Euler, Ritz, and Galerkin to modern computing. SIAM Rev 54, 627–666 (2012)

[19] Gelcich, E.: Geschichte der Uhrmacherkunst. Reprint, Zentralantiquariat der DDR, Weimar (1892)

[20] Geradin, M., Rixen, D.: Mechanical Vibrations. Wiley, New York (1997)

[21] Goldstein, H., Poole, C., Safko, J.: Classical mechanics, 3rd edn. Addison-Wesley, Reading (2001)
[22] Greiner, W.: Classical mechanics: systems of particles and Hamiltonian dynamics. Springer, Berlin (2008)
[23] Gross, D., Hauger, W., Schröder, J., Wall, W., Govindjee, S.: Engineering mechanics 3, 2nd edn. Springer, Berlin (2014)
[24] Guckenheimer, J., Holmes, P.: Nonlinear oscillations, dynamical systems, and bifurcations of vector fields. In: Applied Mathematical Sciences, 7th edn., vol. 42, Springer, New York (2002)
[25] Hadwich, V.: Modellbildung in mechatronischen Systemen. Fortschritt-Berichte VDI, VDI Verlag (1998)
[26] Hagedorn, P.: Non-linear Oscillations. Oxford University Press, New York (1981)
[27] Hahn, W.: Stability of Motion. Springer, Berlin (1967)
[28] Hamel, G.: Theoretische Mechanik. Springer, Berlin (1978)
[29] Huber, R., Clauberg, J., Ulbrich, H.: Herbie: Demonstration of gyroscopic effects by means of a RC vehicle. In: Proceedings of the ASME 2011 IDETC/CIE Conference, Washington, August 29-31 (2011)
[30] Ibrahimbegovic, A.: On the choice of finite rotation parameters. Comput. Meth. Appl. Mech. Eng. 149, 49–71 (1997)
[31] Kane, T.: Mechanical demonstration of mathematical stability and instability. Int. J. Mech. Eng. Educ. 2(4), 45–47 (1974)
[32] von Karman, T.: Die Wirbelstraße. Hoffmann und Campe, Hamburg (1968)
[33] Kauderer, H.: Nichtlineare Mechanik. Springer, Berlin (1958)
[34] Kirchhoff, G.: Vorlesungen über Mathematische Physik, Mechanik. Teubner, Leipzig (1876)
[35] Kücükay, F.: Dynamik der Zahnradgetriebe, Modelle, Verfahren, Verhalten. Springer, Berlin (1987)
[36] La Salle, J., Lefschetz, S.: Stability by Liapunov's direct method with applications. Academic Press, New York (1961)
[37] Laursen, T.: Computational contact and impact mechanics. Springer (2002)
[38] Leipholz, H.: Stability Theory. Wiley, New York (1987)
[39] Lichtenberg, A., Lieberman, M.: Regular and Stochastic Motion. Springer, Berlin (1983)
[40] Magnus, K.: Kreisel-Theorie und Anwendungen. Springer, Berlin (1971)
[41] Magnus, K., Müller-Slany, H.: Grundlagen der technischen Mechanik, 7th edn. Leitfäden der angewandten Mathematik und Mechanik, Teubner, Wiesbaden (2005)
[42] Magnus, K., Popp, K., Sextro, W.: Schwingungen, 8th edn. Vieweg, Wiesbaden (2008)
[43] Meirovitch, L.: Analytical Methods in Vibrations. The MacMillan Company, New York (1967)
[44] Minorsky, N.: Nonlinear Oscillations. Krieger Publishing Company, Princeton (1974)
[45] Müller, P.: Stabilität und Matrizen. Springer, Berlin (1977)
[46] Müller, P., Schiehlen, W.: Lineare Schwingungen. Koch Buchverlag, Planegg (1982)
[47] Nayfeh, A.: Problems in Perturbation. Wiley, New York (1993)
[48] Nayfeh, A., Balachandran, B.: Applied Nonlinear Dynamics. Wiley, New York (1995)
[49] Papastavridis, J.: Tensor calculus and analytical dynamics. Taylor & Francis, London (1998)
[50] Papastavridis, J.: Analytical Mechanics: A Comprehensive Treatise on the Dynamics of Constrained Systems: For Engineers, Physicists, and Mathematicians. Oxford University Press (2002)

[51] Pfeiffer, F.: Mechanical System Dynamics. Springer, Heidelberg (2008)
[52] Pfeiffer, F., Fritzer, A.: Resonanz und Tilgung bei spielbehafteten Systemen. J. Appl. Math. Mech. 4, 38–40 (1992)
[53] Pfeiffer, F., Glocker, C.: Multibody Dynamics with Unilateral Contacts. Wiley, New York (1996)
[54] Popper, K.: Objektive Erkenntnis, ein evolutionärer Entwurf. Hoffmann und Campe, Hamburg (1993)
[55] Post, J.: Objektorientierte Softwareentwicklung zur Simulation von Antriebssträngen, Fortschritt-Berichte VDI: Reihe 11, vol 317. VDI Verlag, Düsseldorf (2003)
[56] Rayleigh, J.: The Theory of Sound. Macmillan, London (1877)
[57] Ritz, W.: über eine neue Methode zur Lösung gewisser Variationsprobleme der mathematischen Physik. Journal für die reine und angewandte Mathematik 135:1–135:61 (1908)
[58] Rudin, W.: Principles of mathematical analysis, 3rd edn. International series in pure and applied mathematics, McGraw-Hill, New York (2008)
[59] Schiehlen, W., Eberhard, P.: Technische Dynamik, 3rd edn. Teubner, Wiesbaden (2012)
[60] Schlichting, H.: Grenzschichttheorie. Springer, Berlin (2006)
[61] Schmidt, G.: Parametererregte Schwingungen. VEB Deutscher Verlag der Wissenschaften, Berlin (1975)
[62] Schwertassek, R., Wallrapp, O.: Dynamik flexibler Mehrkörpersysteme. Vieweg, Wiesbaden (1999)
[63] Shabana, A.: Dynamics of multibody systems, 3rd edn. Cambridge University Press, New York (2005)
[64] Stoer, J., Bulirsch, R.: Introduction to numerical analysis. Texts in applied mathematics, vol. 12. Springer, New York (2010)
[65] Strang, G.: Introduction to Linear Algebra. Wellesley-Cambridge Press, Wellesley (2009)
[66] Synge, J.L.: Classical Dynamics, in Encyclopedia of Physics, Volume III/1: Principles of Classical Mechanics and Field Theory. Springer, Berlin (1960)
[67] Szabo, I.: Geschichte der mechanischen Prinzipien und ihrer wichtigsten Anwendungen, Korr. Nachdruck der 3. Auflage edn. Birkhäuser, Basel (1996)
[68] Szabo, I.: Einführung in die Technische Mechanik, 8th edn. Springer, Berlin (2003)
[69] Tauchert, T.: Energy Principles in Structural Mechanics. McGraw-Hill, New York (1974)
[70] Thom, R.: Structural Stability and Morphogenesis. Westview Press, Boulder (1994)
[71] Thompson, M., Stewart, B.: Nonlinear Dynamics and Chaos. Wiley, New York (2001)
[72] Ulbrich, H.: Dynamik und Regelung von Rotorsystemen, Fortschritt-Berichte VDI: Reihe 11, vol 86. VDI Verlag, Düsseldorf (1986)
[73] Ulbrich, H.: Maschinendynamik. Teubner Studienbücher, Teubner, Stuttgart (1996)
[74] Wriggers, P.: Computational contact mechanics, 1st edn. Wiley, Chichester (2002)
[75] Zeemann, E.: Catastrophe Theory, Selected Papers 1972-1977. Addison-Wesley, Reading (1977)
[76] Zeidler, E., Hackbusch, W., Schwarz, H.R.: Oxford Users' Guide to Mathematics. Oxford University Press, New York (2004)
[77] Ziegler, F.: Mechanics of solids and fluids, 2nd edn. Springer, Berlin (1998)
[78] Zienkiewicz, O., Taylor, R., Zhu, J.: The Finite Element Method Set, 6th edn. Butterworth-Heinemann, Oxford (2005)

Index

Printed in the United States
By Bookmasters